**SYSTEMATIC INDEX OF
RECENT AND PLEISTOCENE
PLANKTONIC FORAMINIFERA**

SYSTEMATIC INDEX OF RECENT AND PLEISTOCENE PLANKTONIC FORAMINIFERA

by
Tsunemasa Saito
Yamagata University

Peter R. Thompson and Dee Breger
Lamont-Doherty Geological Observatory, Columbia University

UNIVERSITY OF TOKYO PRESS

Supported in part by the Ministry of Education, Science and Culture
under a Publication Grant-in-Aid.

CONTENTS

ACKNOWLEDGMENTS

All the scanning electron microscopic work and most of the taxonomic studies which form the basis of this Systematic Index were accomplished at the Lamont-Doherty Geological Observatory of Columbia University, Palisades, New York, while the senior author was serving as Senior Research Associate of the Observatory.

This work was originally conceived to provide a taxonomic guide for a large number of researchers working under the CLIMAP (Climate: Long-range Investigation Mapping and Prediction) project, who would produce data on the present-day seabed distribution of planktonic foraminifera. These distributional data were in turn used to develop theoretical ecological models, for understanding those distributions and the models thus developed were used to interpret the paleogeographic distribution of their fossil counterparts.

For the greatest accuracy of presentation of species concept, all illustrated material was taken from samples as close as possible to the originally described locality or geologic section. Although a greater number of specimens was selected in so far as possible from material in the collections at Lamont-Doherty Observatory, we are indebted to many authors who have proposed new taxa in recent years for their generous donation of topotypic specimens.

The present work received support from the U.S. National Science Foundation under the CLIMAP project through grants GX72-28671 and GA-19690. Sediment cores from which the illustrated specimens were obtained were collected and curated under Office of Naval Research grant N00014-C-0210 and NSF grant OCE 76-18049. A Grant-in-Aid for Scientific Research from the Ministry of Education, the Government of Japan, also aided part of the research done by T. SAITO.

INTRODUCTION

An increasing trend towards inter-institutional cooperation on global synoptic studies necessitates a standardization of microfossil taxonomy for both communication within the group and presentation of unified data to the general scientific community. The limits of desirable taxonomic subdivision depend largely upon the needs of the particular investigation. For example, those involved in a project to create paleoenvironmental reconstructions may be more interested in delimiting community controls required by dominant taxa represented only by the major genera or higher taxonomic catagories; on the other hand, biostratigraphers concerned with time-related phylogenetic development might wish to recognize as many species and varieties as possible. One main concern of both viewpoints is the need for rigorous and correct separation or inclusion of the individual specimens or populations under the valid available nomenclature. However, access to adequate or up-to-date documentation of individual taxa either from the literature or from identified samples is often difficult, especially for non-paleontologists and students new to the field.

Historically, the optical microscope has been the main tool in the study of foraminifera. Early workers sketched their specimens either directly from their microscopic observations or used a camera lucida, adding details they noted or felt ought to be there. As a direct result, new taxa seem to have been recorded when even small variations from established species were noticed. Due to the limited literature available, many oceanographic expeditions initially repeated taxonomic identification, and discovered the duplication later. Also, many published accounts merely provide species descriptions without accompanying figures (*"nomen obscurum"*) (*e.g.,* EHRENBERG's 1845 papers) or make references to species without completing morphological descriptions of their taxonomic status (*"nomen nudum"*). Many illustrated species are clearly synonymous with our present-day concepts of certain species, while others are too poorly illustrated and the collections of original materials have been lost (*"nomen dubium"*). Clearly, the historical validity of many names needs to be carefully studied in the manner of BANNER and BLOW (1960 a, b), with old collections restudied and the International Code of Zoological Nomenclature fully utilized. The Index for Taxonomic Changes lists a considerable number of species, mostly from EHRENBERG and d'ORBIGNY, which may actually have taxonomic priority over existing names. Some undoubtedly represent benthic taxa and others probably should be suppressed (*"nomen oblitum"*), but many may be quite valid.

Development of A Classification Based on Surface Ultramicrostructure

The underlying purpose of any classification is to "pigeon-hole" objects into some sort of hierarchical system, which presumably will reflect the relation of each obejct to the others. In dealing with biota, our aim is to ultimately discern phylogenetic evolution in the group of organisms being studied. In the pioneering stage when little is known about the reproduction and ecology of the organisms, artificial classifications can be erected on the basis of unifying and differentiating morphological features. As more detailed studies on ontogeny and population structure become available they often reveal new insights which require revision, reclassification and subdivision. The older notions, while certainly not wrong *per se,* must be updated.

Prior to the advent of SEM (Scanning Electron Microscope) techniques, the family-level classification of foraminifera (Bolli, Loeblich and Tappan, 1957, p. 21; Loeblich and Tappan, 1964, p. C153) was based on "the wall composition and structure, general chamber arrangement (*i. e.,* type of coiling), basic position of primary aperture (in adults of simpler forms, in the ontogeny of specialized forms)." Banner and Blow (1959, p. 2) based theirs on stratigraphically restricted "external structural modifications of the apertures." Parker (1962), however, proposed a markedly different classification of planktonic foraminifera based on the presence or absence of spines on the living animal. Lipps (1966) suggested a scheme based on several types of wall structure.

The use of the SEM has provided new directions to microfossil study by permitting detailed observation of spines and test surfaces and other structures too minute to study adequately under optical microscopy. Features such as pore-pits, triradiate spines, cancellate surfaces and multiple pustules seem to be related to the planktonic mode of life pursued by pelagic foraminifera (Saito, Thompson and Breger, 1976; Bé *et al.,* 1980). Work by Bé and his co-workers (*e. g.,* Bé and Anderson, 1976; Spindler *et al.,* 1979) on living species of *H. pelagica* shows that the spines support much of the protoplasmic capsule, and that symbiotic algae pass in and out of the host through the pores and along the pseudopodia. Anderson and Bé (1976a) has also noted that different species of foraminifera host different algae; this symbiotic relationship, when pushed to its extreme, could conceivably one day be used to identify genera and even species. It is easy to trace further the ultramicrostructural features of dated species to their ancestral forms.

Once researchers are satisfied with the internal consistency and naturalness of a new biological classification technique, they soon feel compelled to begin using it to name things, usually from the family level on down, creating genera, subgenera, species, subspecies and varieties to accommodate all the taxa recognized. Admittedly, we have also done this to a limited extent, but we have been reluctant to imply more than very general family ties to species without thoroughly studying ancestral relationships.

In a previous paper (Saito, Thompson and Breger, 1976), we have shown that ultra-microstructures of foraminiferal tests, particularly the presence or absence of spines and spine morphology, provide a most useful criterion to establish the natural relationships of foraminiferal taxa. On this basis, we have shown that the majority of Late Cenozoic planktonic foraminifera can be classified into three families: Globigerinidae—tests bearing spines having round or round-becoming triradiate cross sections; Hastigerinidae—tests bearing spines with triradiate cross sections and a row of barbs on the spine blades; and Globorotaliidae—tests barren of spines but having a distinctly pitted surface. Because of this observation, all the species discussed in this Systematic Index are accompanied by ultramicrostructural observations of their test surface as an additional aid in species identification.

At this point, two important concepts need to be clarified: "spinosity" *versus* "pustu-losity" and "secondary shell cortex thickening."

Many workers have noted the occurrence of either single, clumped or multiple crystallites on and protruding out of the test surface. These protrusions have variously been termed spines, pustules, pseudospines, etc., often without making the necessary distinction, although some distinctions have been made (CUSHMAN, 1927, *et seq.*). A spine, as used here, is understood to be a single crystal of calcite elongated along the c-axis (WOOD, 1949, p. 240).

From the work of HEMLEBEN (1969, 1971) and SAITO, THOMPSON and BREGER (1976), it has been shown that planktonic foraminiferal spines are not directly attached to the test, but rather fit into a socket created on top of the spine base (Figs. 1–3). In the living animal it has been shown (BÉ and ANDERSON, 1976, *Science,* no. 4242, cover) that spines provide a "tent-pole-like" support for the living organism's capsule and give the pseudopodia, with their symbiotic algae, paths to travel in and out of the solid shell. According to ANDERSON and BÉ (1976b) the spines of some species are quite flexible, but additionally, and perhaps more significantly, the spines are shed by at least some taxa at the time of reproduction, perhaps to facilitate zygote movement (BÉ, 1980).

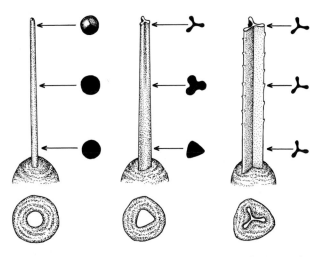

Fig. 1 Spines and spine bases. Left, *Globigerina*-type spine having a round spine base and round spine; middle, *Globigerinella*-type spine having a rounded-triangular spine base and a round or rounded-triangular spine becoming a triradiate spine with growth; right, *Hastigerina*-type spine having triradiate spine base and barbed, triradiate spine.

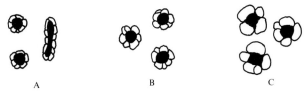

A   B   C

Fig. 2 Petal-shaped calcite crystals developed around the pore-opening of the genus *Globigerinita* (A) and *Candeina* (B, C).

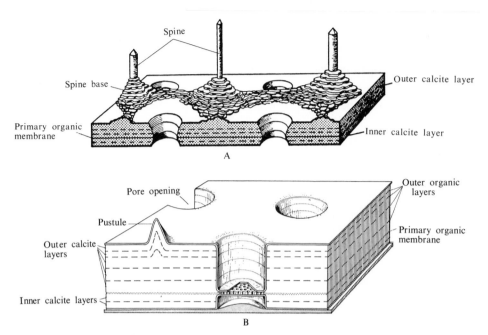

Fig. 3 Schematic cross-section of chamber wall of planktonic foraminifera. A : Spinose species (after HEMLEBEN, 1969 ; with some modifications). B : Non-spinose species (after HEMLBEN, BÉ, ANDERSON and TUNTIVATE, 1977).

Even if people are not convinced by the argument of ALLEE and SCHMIDT (1951, p. 272) that spines are the mechanism which retards sinking in the water, these important physiological characteristics cannot be overlooked. At present, pustules have not been proven to have any vital function, although they may be structures analogous to the spines. They rarely, if ever, occur on spinose taxa, or *vice versa*. SEM studies by us and others have yet to disclose any exact correspondence between spines and pustules. A study by BURT and SCOTT (1975) illustrates the frustration of regarding spines and pustules as comparable structures in the case of *Pulleniatina*. Spine bases, on either early or late chambers, cannot be confused with pustules, whether the pustules are hollow or not. Enough workers have studied all stages of planktonic foraminiferal growth so that spinose specimens referable to *Pulleniatina* would have been seen if they do exist. BURT and SCOTT (1975) have suggested a resorption mechanism for the elimination of spines (*sic*) in the later ontogeny, along with a change in the coiling mode and changes in the outer shell covering.

This sort of discussion leads directly into considerations of the observed heavy outer test wall in advanced stages of ontogeny. A controversy that has existed for many years concerns both the nature of this mechanism and its biostratigraphic implications. BÉ and his co-workers (*e.g.*, BÉ, 1965; BÉ and HEMLEBEN, 1970) have demonstrated that the organism deposits a "secondary cortex" with maturity, much thicker than that which it possesses in juvenile stages. Observing the same species, *Sphaeroidinella dehiscens*, BLOW (1969) has strongly disagreed with BÉ by pointing out that *Globigerinoides trilobus* and *G. sacculifer* have separate phylogenies from that of *Sphaerodinella dehiscens* and cannot be conspecific. A "stalemate" exists that must be resolved (for further discussion see BLOW, 1979, pp. 636, 728). Specimens may be found to illustrate both viewpoints, although it must be conceded that *S. dehiscens* is much rarer in deep-sea sediments than *G. sacculifer*.

Many species have been created based on shell-thickening, most recently *Globorotalia truncatulinoides pachytheca* BLOW and *Globorotalia tosaensis tenuitheca* BLOW. As admitted by BLOW (1969), these subspecies are differentiated from the typical form only by their thicker shell and some morphological modifications accompanying this. Beyond this these species serve no practical purpose but add taxa where other information is perhaps more desirable.

Again we return to the philosophy behind the creation or suppression of species. It is always desirable to differentiate index forms and environmentally restricted forms. But is a plethora of species really necessary or even desirable, or will a few expanded population concepts be sufficient for practical purposes? To criticize effectively, one must provide an alternative. We do not disparage the work of the past or even the intentions of that work, but rather we wish to update it in light of more recent developments. Population concepts (SCOTT, 1974) need to be broadened to recognize that every specimen does not look exactly like every other one. BERGER (1969, 1971) pointed out that many taxonomic concepts are founded on abnormal or atypical specimens (kummerforms), with additional forms representing normal growth (see also AKTÜRK, 1976). It also must be noted that ecology plays a major role in directing physiological responses, producing many similar-looking forms in upper waters, where controls are basically the same (BURT and SCOTT, 1975). Little attention has been paid to variations at biogeographic margins, where stresses may be more severe than in the population center. The forms from these marginal regions cannot be expected to resemble those from other regions, although we may be dealing with exactly the same organism. KENNETT (1968a, b) has ably shown the latitudinal variation of *G. truncatulinoides* and *N. pachyderma*. BANDY (1972) reported on the variations of *G. bulloides*. A series of papers by MALMGREN and KENNETT (1972, 1976, 1977) applied multivariate statistical analyses of morphological variations to *N. pachyderma, G. bulloides* and *G. falconensis* to examine if any of these species having a rather simply built test would exhibit a clinal morphologic variation in response to changing environmental conditions such as a climatic gradient. A thorough review by KENNETT (1976) of many articles discussing the problem of phenotypic variation in Late Cenozoic planktonic foraminifera reveals that such morphologic characters as coiling direction, test size, shape of final chamber, test porosities, and test thickness all do show a positive correlation with changing oceanographic conditions, notably with temperature. Similar studies could be done on every taxa, and SEM study can be expected to facilitate the development of sound conclusions.

In conclusion, our approach to the great number of proposed species is a conservative one. The combination of the apparent great tolerability shown by planktonic foraminifera and the equally variable structure of ocean currents, temperatures and salinities suggests the likelihood that many separate "species" are ecological variants of other taxa. While these "formae" or "phenotypes" have ecological usefulness, they should be carefully reconsidered before being assigned to independent taxonomic categories.

HOW TO USE THIS GUIDE

Five families and twenty genera of Pleistocene to Recent planktonic foraminifera are recognized in this paper. Fig. 4 shows a suggested phylogenetic radiation scheme for the derivation of these modern genera from various Tertiary stocks. Of the five families, two are spinose and three are not: Hastigerinidae (triradiate, barbed spines); Globigerinidae (round or round-becoming-triradiate spines); Candeinidae (non-spinose with infralaminal apertures); Globorotaliidae (non-spionse, but having pustulate and/or pitted surface); and Heterohelicidae (non-spinose biserial forms represented only by one genus and one species). The scanning electron micrographs (SEM) of the species in this paper have been arranged according to the family and genera arrangement shown in Fig. 4.

One does not always know in advance whether he is looking at a *Globigerina* or a *Globorotalia,* but with a light microscope and a little patience they can be distinguished. In this regard, we have devised two approaches to facilitate identification: a species key (p. 21) and a guide to synonyms (p. 166). The guide to synonyms is an alphabetic compilation of all the synonyms recognized in this volume, arranged by species' epithet. Any literature identification can be "translated" into the taxonomy employed here (barring any misidentification in the paper being "translated").

At the other extreme is the identification of an unknown specimen using the key provided. Although based on SEM study, every effort has been made to use light microscope conventions of terminology such as spinosity, rugosity, shininess, chamber morphology, etc. Figs. 1–3 illustrate the ultramicrostructure terminology of the key; basic morphologic terms are in the same sense of BOLLI, LOEBLICH and TAPPAN (1957). From first impressions of the most noticeable features, the key gradually leads to the level of recognition of several possible species. At this point, one is referred to the appropriate plates of the most similar-looking taxa with accompanying diagnoses and remarks which should finalize identification. An obvious drawback, a direct complication of space limitations, is that it is impossible to figure more than a very few specimens of each species. Two compensations have been attempted. First, the synonymy of each species includes not only the major name changes, but also additional papers with special treatments or high-quality illustrations. Second, every effort has been expended to locate material as similar to the original intent as possible. Section 5 (p. 22) explains the hierarchy of sample-type preference (after BLACKWELDER, 1967). In this way, the worker can gauge the morphological characteristics of the specimens figured here in comparison to the original and whatever material one is examining.

In long-term practice, the routine examination of numerous specimens from many samples will serve to illustrate the variability of the species more than the few shown here. It is hoped that the variability within the population structure under widely variable environments will suggest new courses of study and improvements on this system. At such a point, the taxonomy for each taxa and the bibliography presented here should be invaluable.

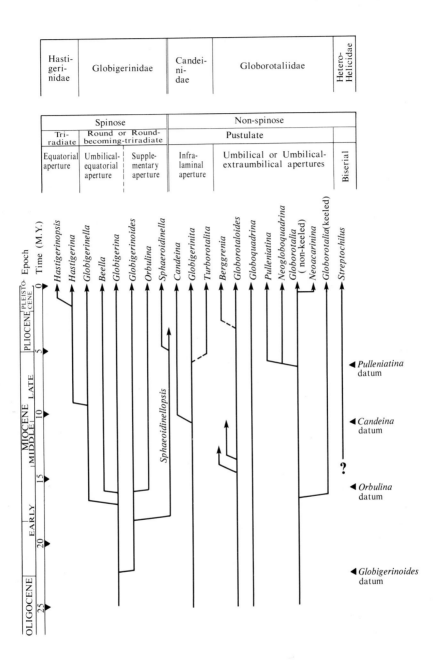

Fig. 4 Phylogenetic derivation and stratigraphic ranges of modern genera of planktonic foraminifera with their suggested familial classification based on the ultramicrostructures of test walls.

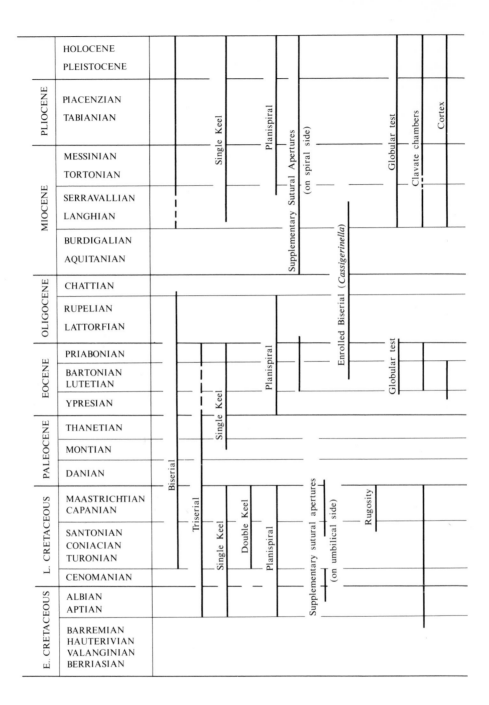

Fig. 5 Stratigraphic distribution of some major test modifications and coiling morhotypes observed in planktonic foraminifera. Evolution of homeomorphic morphotypes in phylogenetically unrelated species have been repeated during the Cretaceous-Cenozoic interval as much as three times for certain morphotypes.

1. SEM Preparation

To produce the scanning electron micrographs presented here, the foraminiferal specimens were prepared by the following methods: for routine specimens, after a standard aluminum stub was briefly rinsed in acetone and dried, a small square of double-stick tape was fixed to the top center portion of the stub and a line of butyl-acetate silver-based paint was drawn around the tape's sides, connecting the top surface with the stub to provide electrical conductivity. The stubs thus prepared were degassed for two hours in a Denton Vacuum Evaporator. Under a light microscope, cleaned and acetone-rinsed foraminiferal specimens were then mounted on the tape with a fine picking brush moistened slightly with distilled water. To prepare particularly minute species such as *Globigerinita uvula,* a solution of gum tragacanth well diluted with distilled water was brushed directly over the stub's entire top surface and allowed to dry. The moist picking brush used for transferring the specimen onto the stub normally carried enough water to effect a bond between the specimen and the stub. For preparing the more smooth-surface textured species such as *Pulleniatina finalis,* a thin layer of butyl-acetate silver-based paint was applied directly onto the stub's top surface and the specimens quickly positioned before the brush stroke dried; this method provides a better electrical ground for such species. In all techniques, when three views of one specimen were desired, the specimen was positioned with the umbilical and spiral views exposed at an angle of 90° from the horizontal, which allowed for these two and the side view as well to be recorded. In those instances when a specimen was to be used for one view (umbilical or spiral) or two (the second always being the side view), it was mounted with the desired umbilical or spiral face exposed at 70° from the horizontal. Up to twelve mounted stubs could be placed in the rotary stage of the vacuum evaporator at one time. At a vacuum of 3×10^{-5} Torr, eight to twelve inches (depending on the humidity) of 8-mil gold wire were gently evaporated from each of two tungsten filaments located at 5° and 70° from the surface of the rotating stage. The desired views of specimens prepared in these ways were photographed in a Cambridge Scanning Electron Microscope with an Exacta 500 camera using Kodak 35 mm, Panatomic-X fine-grain film. The micrographs were printed to standardized magnifications for each plate by varying the degree of photographic enlargement. Some "dodging" and "burning in" was done on most prints to present them at the best advantage. Retouching was done where necessary to mute occasional "holidays" or artifacts or to minimize defects such as scratches on the negatives.

2. Index for Taxonomic Changes of Planktonic Foraminifera (?) listed as Quaternary

?Aristerospira alloderma EHRENBERG, 1858, p. 15 (no figures). — EHRENBERG, 1857, pp. 546,548,550,552,555 (*nomen nudum*). — EHRENBERG, 1873, p. 221, pl. 11, fig. 12. Maris Cretici Deep, Aegean Sea, 9,720 ft.

?Aristerospira angustior EHRENBERG, 1873, pp. 181,221, Table 3a, opp. p. 173, pl. 12, fig. 8. Caspian Sea, 90 ft.

Aristerospira bacheana EHRENBERG, 1861, pp. 225,228,302. — EHRENBERG, 1873, p. 221, pl. 5, fig. 15 (given in plate caption as *A. bacheiana*). Gulf of Mexico (Lat. 28° 29′N, Long. 87°01′30″W, 2,556 ft).

Aristerospira bakuana EHRENBERG, 1873, pp. 181,221, Table 1c, opp. p. 171, pl. 12, fig. 3. Rhede von Baku, Caspian Sea.

Aristeropira bupthalma EHRENBERG, 1873, pp. 181, 221, Table 3c, opp. p. 171, pl. 12, fig. 5. Rhede von Baku, Caspian Sea.

Aristerospira crassa EHRENBERG, 1872, p. 278. — EHRENBERG, 1873, p. 221, pl. 3, fig. 9. America-England telegraph line (Lat. 51° 15′N, Long. 34° 16′W, 9,780 ft).

Aristerospira cucullaris EHRENBERG, 1872, p. 278. — EHRENBERG, 1874, p. 463, pl. 1, figs. 10,11. Greenland Sea, Arctic Expedition of Germaniae II (1869–1870) (Lat. 71° 37′ N, Long. 12° 23′W, 7,542 ft).

Aristerospira globularia EHRENBERG, 1858, p. 16. — EHRENBERG, 1873, p. 221, pl. 11, fig. 11. Maris Cretici Deep, Aegean Sea, 9,720 ft.

Aristerospira glomerata EHRENBERG, 1861, pp. 276,277,303. — EHRENBERG, 1873, p. 221, pl. 1, fig. 5. Greenland Deep, Davis Strait (Lat. 62° 40′N, Long. 20° W, 6,000 ft).

Aristerospira integra EHRENBERG, 1873, p. 221, pl. 5, fig. 13. Lat. 28° 29′N, Long. 87° 01′30″W, 2,556 ft.

Aristerospira isoderma EHRENBERG, 1858, p. 16. — EHRENBERG, 1873, p. 221, pl. 11, fig. 9. — EHRENBERG, 1857, pp. 548,552,555 (*nomen nudum*). Maris Cretici, Aegean Sea, 9,720 ft.

Aristerospira laevigata EHRENBERG, 1858, p. 16. — EHRENBERG, 1873, p. 221, pl. 11, fig. 10. — EHRENBERG, 1875, pp. 552,556 (*nomen nudum*). Maris Cretici, Aegean Sea, 9,720 ft.

Aristerospira lepida EHRENBERG, 1872, p. 279. — EHRENBERG, 1873, p. 221, pl. 3, fig. 5. America-England telegraph line (Lat. 51° 15′N, Long. 34° 16′W, 9,780 ft).

Aristerospira liopentas EHRENBERG, 1861, pp. 276,277,303. — EHRENBERG, 1873, p. 221, pl. 1, fig. 1. Greenland Deep (Lat. 62° 49′N, Long. 29° W, 6,000 ft).

Aristerospira major EHRENBERG, 1873, p. 221, Table 3c, opp. p. 173, pl. 12, fig. 2. Caspian Sea, 90 ft.

Aristerospira mauryana EHRENBERG, 1861, pp. 225,229,303. — EHRENBERG, 1873, p. 221, pl. 5, fig. 5. Gulf Stream near Florida (Lat. 27° 54′N, Long. 86° 05′W, 9,066 ft).

Aristerospira microtetras EHRENBERG, 1861, pp. 276,278,303. — EHRENBERG, 1873, p. 222, pl. 1, fig. 2. Greenland Deep (Lat. 59° 12′N, Long. 50° 38′W, 10,998 ft) and Davis Strait (Lat. 62° 40′N, 29° W, 6,000 ft).

Aristerospira omphalotetras EHRENBERG, 1872, p. 279 — EHRENBERG, 1873, p. 223, pl. 3, fig. 15. Lat. 51° 15′N, Long. 34° 16′W, 9,780 ft; Lat. 52° 26′N, Long. 26° 30′W, 11,580 ft.

Aristerospira platypora EHRENBERG, 1858, p. 17 — EHRENBERG, 1857, pp. 548,556 (*nomen nudum*). — EHRENBERG, 1873, p. 223, pl. 11, fig. 20. Maris Cretici, Aegean Sea, 3,000 ft.

Aristerospira porosa EHRENBERG, 1861, pp. 224,225,276,303. — EHRENBERG, 1873, p. 223, pl. 1, fig. 3. Lat. 24° 23′N, Long. 80° 43′W, 630 ft; Lat. 28° 12′N, Long. 85° 47′W, 960 ft; Lat. 25° 37′40″ N, Long. 80° 01′05″W, 1,158 ft; Lat. 28° 28′N, Long. 87° 01′30″W, 2,556 ft; Lat. 27° 54′N, Long. 86° 05′W, 9,066 ft; Lat. 54° –62° 04′ N, Long. 29° –51° 50′W, 300–12,540 ft.

Aristerospira schaffneri EHRENBERG, 1872, p. 278. — EHRENBERG, 1873, p. 223, pl. 4, fig. 16. Lat. 52°03′N, Long. 15° 02′W, 2,460 ft.

Aristerospira scutata EHRENBERG, 1861, pp. 224,225,303,308. — EHRENBERG, 1873, p. 223, pl. 5, fig. 4. Lats. 24° 28′–31° 01′49″N, Longs. 79° 08′30″–87° 01′30″W, 69–9,066 ft.

Globigerina adriatica FORNASINI, 1898, p. 582, pl. 3, figs. 5–7. Syntype locality—Porto Corsins, Ravenna, Italy.

Globigerina alloderma EHRENBERG, 1873, p. 225, pl. 5, fig. 7. Lat. 27° 54′N, Long. 86° 05′W, 9,066 ft.

Globigerina detrita TERQUEM, 1875, p. 20. — TERQUEM, 1876, p. 435, pl. 4, figs. 4a–d. Syntype locality—Plage de Dunkerque, Department du Nord, France.

Globigerina globularis D'ORBIGNY, 1826, p. 277, fig. 3 (*nomen nudum*). — FORNASINI, 1903, p. 140, pl. 1, figs. 1, 1a-b. Syntype locality—near Mauritius, Indian Ocean.

Globigerina helicina D'ORBIGNY, 1826, p. 277, list no. 5 (no figures). Recent deposits of the Adriatic Sea, near Rimini. — BANNER and BLOW, 1960a, pp. 13–14, pl. 2, figs. 5a–c (lectotype).

Globigerina juvenilis BOLLI. — BERMÚDEZ and BOLLI, 1969, pp. 152–153, pl. 2, figs. 10–12.

Globigerina nereidum EHRENBERG, 1872, p. 282 (no figures). Lat. 50° 40'N, Long. 37° 15'W, 9,600 ft.

Globigerina quadrangularis RHUMBLER, 1911. — ANONYMOUS, 1949, pl. 30, figs. 18–21. = *Sphaeroidinellopsis quadrangularis* (RHUMBLER). — BERMÚDEZ, 1961. Atlantic Ocean.

Globigerina radians EGGER, 1893, p. 362, pl. 13, figs. 22–24. Syntype localities — Station 87 (Lats. 20° 49'–20° 41'S, Longs. 113° 46'–114° 17'E, 915 m); Station 90 (Lat. 18° 52'S, Long. 116° 18'E, 359 m); Station 130 (Lat. 14° 52.4'S, Long. 175° 32.7'W, 1,655 m).

Globigerina regularis TERQUEM, 1880, p. 49, pl. 16, figs. 2a–b. Syntype locality — Plage du Dunkerque, Department du Nord, France.

Globigerina sphaeroides EGGER, 1893, p. 367, pl. 13, figs. 69–70. Indian Ocean off Australia and west coast of Africa.

Globigerina ternata EHRENBERG, 1854, p. 247, pl. 35B (group B4), figs. 5–6. Lat. 37° 05'N, Long. 14° 30'W, 840 ft. — RHUMBLER, 1911, pp. 148,217, pl. 29, figs. 5–13.

Globigerina trilocularis D'ORBIGNY. — BERMÚDEZ and BOLLI, 1969, p. 156, pl. 4, figs. 7–9.

Globigerina turkomanica BRODSKY, 1929, p. 22, pl. 1, figs. 10a–b. Salt Wells at Khalka, about 40 km northeast of Yerbent, Kara Kum Desert, Turkman SSR, 7.0 m.

Globorotalia obesa BOLLI. — BERMÚDEZ and BOLLI, 1969, p. 175, pl. 14, figs. 7–12.

Phanerostomum alloderma EHRENBERG, 1861, pp. 276,278,306 — EHRENBERG, 1873, p. 235, pl. 1, fig. 12. Greenland Deep (Lat. 59° 12'N, Long. 50° 38'W, 10,998 m).

Phanerostomum atlanticum EHRENBERG, 1854a, p. 248. — EHRENBERG, 1854b, p. 23, pl. 35B (group 4B), figs. 3–4. Lat. 42° 47'N, Long. 29° W, 6,480 ft.

Phanerostomum globularis EHRENBERG, 1861, pp. 276,278,306. — EHRENBERG, 1873, p. 235, pl. 1, fig. 14. Greenland Deep (Lat. 60° 05'N, Long. 50° 27'W, 12,540 ft).

Phanerostomum micromega EHRENBERG, 1861, pp. 276,277,306. — EHRENBERG, 1873, p. 235, pl. 1, fig. 11. Greenland Deep (Lat. 62° 40'N, Long. 29° W, 6,000 ft).

Phanerostomum microporum EHRENBERG, 1861, pp. 276,277,307. — EHRENBERG, p. 235, pl. 1, fig. 9. Greenland Deep (Lat. 62° 40'N, Long, 29° W, 6,000 ft).

Phanerostomum oceanicum EHRENBERG, 1873, pp. 143,235, pl. 3, fig. 10. First Atlantic telegraph line (1856) (Lat. 51° 15'N, Long. 34° 16'W, 9,780 ft).

Phanerostomum pelagicum EHRENBERG, 1872, p. 287. — EHRENBERG, 1873, p. 235, pl. 4, fig. 14. Lat. 52° 03'N, Long. 15° 02'W, 2,460 ft.

Phanerostomum scutellatum EHRENBERG, 1873, p. 235, pl. 1, figs. 13a–b. Davis Strait (Lat. 62° 40'N, Long. 29° W, 6,000 ft).

Planulina abyssicola EHRENBERG, 1861, pp. 276,277,307. — EHRENBERG, 1873, pl. 1, fig. 15. Greenland Deep (Lat. 62° 06'N, Long. 32° 21'W, 9,240 ft).

Planulina? crisiae EHRENBERG, 1858, p. 23. — EHRENBERG, 1873, p. 235, pl. 11, fig. 18. Maris Cretici, Aegean Sea, 9,720 ft.

Planulina? cymodoceae EHRENBERG, 1857 (*nomen nudum*). — EHRENBERG, 1858, p. 24. — EHRENBERG, 1873, p. 235, pl. 11, fig. 19. Maris Cretici, Aegean Sea, 3,000 ft.

Planulina depressa EHRENBERG, 1861, p. 307. — EHRENBERG, 1873, p. 235, pl. 1, fig. 20. Greenland Deep (Lat. 62° 40'N, Long. 29° W, 6,000 ft).

Planulina diaphana EHRENBERG, 1854, p. 133 (*nomen nudum*). — EHRENBERG, 1873, p. 237, pl. 5, fig. 10. Lat. 28° 29'N, Long. 87° 01'30''W, 2,556 ft.

Planulina heterocyclina EHRENBERG, 1872, p. 289. — EHRENBERG, 1873, p. 237, pl. 4, fig. 6. Lat. 57° 45'N, Long. 21° 09'W, 9,540 ft.

Planulina laevigata EHRENBERG, 1861, pp. 276,277,307. — EHRENBERG, 1873, p. 237, pl. 1, fig. 10. Greenland Deep (Lat. 62° 40'N, Long. 29° W, 6,000 ft).

Planulina leptoderma EHRENBERG, 1861, pp. 224,307. — EHRENBERG, 1873, p. 237, pl. 5, fig. 8. Lat. 27° 54'N, Long. 86° 05'W, 9,066 ft.

Planulina mauryana EHRENBERG, 1872, p. 289. — EHRENBERG, 1873, p. 237, pl. 3, fig. 1. Lat. 52° 26'N, Long. 26° 30'W, 11,580 ft.

Planulina megalopentas EHRENBERG, 1872, p. 289. — EHRENBERG, 1873, p. 237, pl. 4, fig. 8. Lat. 50° 40'N, Long. 37° 15'W, 9,600 ft.

Planulina micropentas EHRENBERG, 1872, p. 289. — EHRENBERG, 1873, p. 237, pl. 4, fig. 8. Lat. 50° 40'N,

Long. 37° 15'W, 9,600 ft.

Planulina perihexas EHRENBERG, 1872, p. 290. — EHRENBERG, 1873, p. 239, pl. 3, fig. 13. Lat. 51° 15'N, Long. 34° 16'W, 9,780 ft.

Planulina seriata EHRENBERG, 1873, p. 239, pl. 5, fig. 9. Lat. 27° 54'N, Long. 86° 05'W, 9,066 ft.

Planulina sphaerocharis EHRENBERG, 1872, p. 290. — EHRENBERG, 1873, p. 239, pl. 4, fig. 9. Lat. 52° 03'N, Long. 15° 02'W, 2,460 ft.

Planulina tenuis EHRENBERG, 1873, p. 144, pl. 3, fig. 2. Lat. 51° 15'N, Long. 34° 16'W, 9,780 ft.

Porospira planulina EHRENBERG, 1872, p. 291. — EHRENBERG, 1873, p. 241, pl. 4, fig. 11. America-England telegraph line (Lat. 57° 45'N, Long. 21° 09'W, 9,540 ft).

Porospira septenaria EHRENBERG, 1872, p. 291. — EHRENBERG, 1873, pl. 4, fig. 10. America-England telegraph line (Lat. 57° 45'N, Long. 21° 09'W, 9,540 ft).

Ptygostomum orphei EHRENBERG, 1854a, p. 248. — EHRENBERG, 1854b, pl. 35B (group B4), figs. 1–2. Lat. 37° 05'N, Long. 14° 30'W, 840 ft.

Pylodexia atlantica EHRENBERG, 1872, p. 292. — EHRENBERG, 1873, pl. 4, fig. 2. Lat.52° 03'N, Long.15° 02'W, 2,460 ft.

Pylodexia globigerina EHRENBERG, 1872, p. 292 (*nomen nudum*). — EHRENBERG, 1873, p. 241. Lat. 52° 03'N, Long. 15° 02'W, 2,460 ft.

Pylodexia glomerulus EHRENBERG, 1861, pp. 276,277,308. — EHRENBERG, 1873, pl. 1, fig. 23, p. 241. Lat. 62° 40'N, Long. 29° W, 6,000 ft.

Pylodexia heteropora EHRENBERG, 1872, p. 292 (*nomen nudum*). Lat. 52° 03'N, Long.15° 02'W, 2,460 ft.

Pylodexia megastoma EHRENBERG, 1872, p. 292. — EHRENBERG, 1873, p. 241. Lat. 34° 55'S, Long. 22° 45'E, 450 ft. = *Aristerospira megastoma* EHRENBERG, 1863 (*nomen nudum*).

Pylodexia platytetras EHRENBERG, 1872, p. 289. — EHRENBERG, 1873, p. 241, pl. 3, fig. 14. Lat. 51° 15'N, Long. 34° 16'W, 9,780 ft.

Pylodexia pusilla EHRENBERG, 1873, p. 241. Maris Cretici, Aegean Sea, 9,720 ft. = *Globigerina pusilla* EHRENBERG, 1858, p. 27 (*nomen nudum*).

Pylodexia rubra EHRENBERG, 1873, p. 293 (*nomen nudum*).

Pylodexia tetratrias EHRENBERG, 1858 (*nomen nudum*). — EHRENBERG, 1873, p. 241, pl. 11, fig. 8. Maris Cretici, Aegean Sea, 6,900 ft. = *Globigerina tetratrias* EHRENBERG, 1858, p. 27 (*nomen nudum*).

?Rotalia abyssorum EHRENBERG, 1858, p. 28. — EHRENBERG, 1873, p. 243, pl. 11, fig. 15. Maris Cretici, Aegean Sea, 3,000–9,720 ft.

?Rotalia cochlea EHRENBERG, 1843, pp. 314,398,428, pl. 2, VI, fig. 42. Sea coast of Cuba.

?Rotalia egena EHRENBERG, 1843, pp. 314,398,428, pl. 2, VI, fig. 43. Sea coast of Cuba.

?Rotalia groenlandica EHRENBERG, 1872, p. 293. — EHRENBERG, 1874, p. 463, pl. 1, fig. 15. Greenland Sea, Germaniae I Arctic Expedition (1861). Lat. 75° 57'N, Long. 12° 11'W, 900 ft.

Rotalia phanerostomum EHRENBERG, 1873, p. 245, pl. 5, fig. 11. Sea floor beneath Gulf Stream (Lat. 28° 12'N, Long. 85° 47'W, 960 ft).

Rotalina canariensis D'ORBIGNY, 1839, p. 130, pl. 1, figs. 34–36. — BANNER and BLOW, 1960a, pp. 33–34, pl. 5, fig. 4 (lectotype).

3. Genera of Planktonic Foraminifera (Pleistocene–Recent)
Order Foraminiferida Eichwald, 1830

I. Family Hastigerinidae SAITO and THOMPSON, 1976

Test trochospiral becoming planispiral or streptospiral; chambers ovate to clavate, often bifurcating or trifurcating; wall calcareous, radial in structure, perforated; only distal end of chambers spinose, spines having triradiate cross-section and frequently barbed along spine-blade edges; primary aperture symmetrical, no secondary apertures.

A. *Hastigerina* THOMSON, 1876

Test planispiral; chambers ovate; wall calcareous, radial, perforated; primary aperture symmetrical, frequently having apertural lip and relict apertures; triradiate spines with upward-pointing barbs on spine-blade edges.

B. *Hastigerinopsis* SAITO and THOMPSON, 1976

Test streptospiral; chambers clavate, often bifurcating or trifurcating; wall calcareous, radial, perforated; aperture interiomarginal, extraumbilical-umbilical becoming spiroumbilical or equatorial; triradiate spines with upward-pointing barbs on spine-blade edges.

II. Family Globigerinidae CARPENTER, PARKER and JONES, 1862

Test trochospiral sometimes becoming planispiral or spherical; chambers globular, ovate or radially elongate; wall calcareous, perforated, radial in structure; spinose, spines are round or round-becoming-triradiate in cross-section; primary aperture umbilical or spiroumbilical; may have secondary sutural or aerial apertures.

A. *Globigerina* D'ORBIGNY, 1826

Test trochospiral; chambers globular; wall calcareous, perforated, radial; spinose, spines round in cross-section; primary aperture umbilical; no secondary apertures.

B. *Globigerinoides* CUSHMAN, 1927

Test trochospiral; chambers globular or ovate; wall calcareous, perforated; spinose, spines round in cross-section; primary aperture umbilical; secondary apertures.

C. *Orbulina* D'ORBIGNY, 1839

Test trochospiral becoming spherical; chambers globular; wall calcareous; perforated, radial; spinose, spines round-becoming-triradiate in cross-section; primary aperture in trochoid forms umbilical, secondary apertures sutural; no aperture in spherical form.

D. *Sphaeroidinella* CUSHMAN, 1927

Test trochospiral; chambers globular; wall calcareous, perforate, covered with heavy cortex, possibly spinose in immature stages; primary aperture umbilical, largely obscured by overhanging cortex turning outwards forming chamber flanges; secondary aperture(s) located on spiral side.

E. *Beella* BANNER and BLOW, 1960

Test trochospiral; chambers globular becoming ovate or radially elongate; wall calcareous, perforated, radial; spinose(?); primary aperture spiroumbilicate; no true secondary apertures; multiple pore structures.

F. *Globigerinella* CUSHMAN, 1927

Test trochoid becoming planispiral; chambers globular to ovate; wall calcareous, perforated, radial; spinose, spines having round or round-becoming-triradiate cross-section; primary aperture umbilical; no secondary apertures.

III. Family Candeinidae SAITO and THOMPSON, this volume

Test trochoid, chambers globular, ovate, or radially elongate; wall calcareous, perforate, non-spinose; primary aperture if present umbilical, usually modified with bullae, or test may have only multiple sutural apertures.

A. *Candeina* D'ORBIGNY, 1839

Test trochoid, chambers globular; wall calcareous, finely perforate, non-spinose; no primary aperture visible; multiple sutural openings on all sides.

B. *Globigerinita* BRÖNNIMANN, 1951

Test trochospiral, chambers globular; wall calcareous, finely perforate, non-spinose, finely pustulate; primary aperture umbilical, typically modified by a bulla; secondary apertures uncommon; areal bullae occur.

C. *Turborotalita* BLOW and BANNER, 1962

Test trochospiral, chambers ovate to radially elongate; wall calcareous, perforate, non-spinose, finely pustulate; primary aperture umbilical to extraumbilical, covered with a bulla-like extension of the base of the final chamber across the umbilicus; no secondary apertures.

IV. Family Globorotaliidae CUSHMAN, 1927

Test trochospiral, occasionally becoming planispiral; chamber shape variable, ranging from angular to ovate, axially or radially elongate, occasionally clavate; may have peripheral keel; wall calcareous, perforated, radial in structure; non-spinose, but surface ranges from smooth to pitted or pustulate; primary aperture extraumbilical-umbilical or umbilical; apertural modifications common, such as lips, teeth, umbilical flaps; may have secondary apertures.

    A. *Globorotalia* CUSHMAN, 1927

    Test trochospiral; chamber shape ranging from angular to ovate; often possesses peripheral keel; wall calcareous, perforated, radial; non-spinose, but surface ranges from smooth to pitted or pustulate; primary aperture extraumbilical-umbilical, apertural modifications usually present such as lips or umbilical flaps; no secondary apertures.

    B. *Globorotaloides* BOLLI, 1957

    Test trochospiral; chamber shape ovate to slightly radially elongate; wall calcareous, perforated, radial; non-spinose, but surface pitted to slightly pustulate; primary aperture extraumbilical-umbilical; apertural modifications usually present, such as lips, teeth or umbilical flaps; no secondary apertures.

    C. *Neogloboquadrina* BANDY, FRERICHS and VINCENT, 1967

    Test trochospiral; chamber shape ovate to subglobular; wall calcareous, perforated, radial; non-spinose, but surface coarsely pitted; primary aperture umbilical; apertural modifications typically present, such as teeth; no secondary aperture.

    D. *Pulleniatina* CUSHMAN, 1927

    Test streptospiral, chambers subglobular; wall calcareous, perforate, non-spinose in mature forms, possibly spinose or coarsely pustulate in juveniles, adults have thick cortex; primary aperture interiomarginal umbilical-extraumbilical; no secondary apertures.

    E. *Globoquadrina* FINLAY, 1947

    Test trochospiral, chambers globular to subglobular; wall calcareous, perforate, non-spinose, pustulate; primary aperture umbilical, typically modified with one or more umbilical teeth; no secondary apertures.

    F. *Neoacarinina* THOMPSON, 1973

    Test trochoidal, chambers subglobular; wall calcareous, perforate, non-spinose, densely covered with coarse multiple pustules; primary aperture umbilical-extraumbilical, commonly covered with a bulla; no secondary apertures.

    G. *Berggrenia* PARKER, 1976

    Test trochospiral, chambers subglobular to radially elongate; wall calcareous, perforate, non-spinose, finely pustulate; primary aperture umbilical-extraumbilical, often with a bulla-like extension of the final chamber; no secondary apertures; test size very small.

V. Family Heterohelicidae CUSHMAN, 1927

Test biserial to multiserial, chambers globular to ovate; wall calcareous, perforate, non-spinose, finely pustulate; primary aperture terminal, often raised on a short neck; secondary apertures rare.

    A. *Streptochilus* BRÖNNIMANN and RESIG, 1971

    Test biserial, chambers subglobular; wall calcareous, non-spinose, finely pustulate; primary aperture terminal, often raised on a short neck with a small collar; no secondary apertures; test size very small.

4. Key to The Genera and Species of The Pleistocene–Recent

I. Test surface rough under light microscope (spines, spine bases, granules, pustules)
  A. Spines visible on living or well-preserved specimens; spine bases on most specimens, located in interpore areas [Hastigerinidae, Globigerinidae]
    1) Spines or spine bases restricted to distal ends of chambers [Hastigerinidae]
      a) Globular to subglobular chambers, planispiral coiling [*Hastigerina*]—species included: *H. pelagica, H. parapelagica*
      b) Clavate chambers, streptospiral coiling [*Hastigerinopsis*]—species included: *H. digitiformans*
    2) Spines or spine bases on all portions of test wall [Globigerinidae]
      a) Primary aperture only
      Radially elongate chambers
        Low trochospire [*Globigerinella*]—species included: *G. adamsi, G. aequilateralis, G. calida*
        Medium trochospire or streptospire [*Beella*]—species included: *B. digitata*
      Globular or spherical chambers [*Globigerina*]—species included: *antarctica, atlantisae, bermudezi, bulloides, cariacoensis, decoraperta, egelida, exumbilicata, falconensis, megastoma, polusi, quinqueloba, rubescens, umbilicata*
      b) Supplementary apertures present
      Spherical to subglobular chambers
        Single spherical chamber [*Orbulina*]—species included: *O. universa*
        Subglobular chambers [*Sphaeroidinella*]—species included: *S. dehiscens, S. excavata*
        Subglobular-spherical chambers [*Globigerinoides*]—species included: *G. conglobatus, G. elongatus, G. fistulosus, G. gomitulus, G. obliquus, G. pyramidalis, G. ruber, G. sacculifer, G. tenellus; O. universa* (trochoid form)
  B. Pustules or granules visible under light microscope on test surface, no spines or spine bases [Globorotaliidae]
    1) Surface granular, coarsely pitted
      a) Pustules present only near aperture
      Apertural tooth present [*Globoquadrina*]—species included: *G. conglomerata, G. pseudofoliata*
      No apertural tooth [*Globorotaloides*]—species included: *G. hexagona*
      b) Pustules not prominent [*Neogloboquadrina*]
      Low to medium trochospire—species included: *asanoi, blowi, eggeri, himiensis, humerosa, kagaensis, incompta, cryophila, pachyderma, polusi, pseudopachyderma*
      Medium to high trochospire—species included: *dutertrei*
    2) Surface pustulate
      a) Peripheral keel absent
      Singular pustules [*Globorotalia*]—species included: *crassula, hessi, hirsuta, inflata, neominutissima, oceanica, parkerae, puncticulata, ronda, scitula, seigliei, tosaensis*
      Multiple pustules [*Neoacarinina*]—species included: *N. blowi*
      b) Peripheral keel present [*Globorotalia* (keeled)]—species included: *cavernula, crozetensis, fimbriata, flexuosa, lata, menardii, neoflexuosa, pertenuis, theyeri, truncatulinoides, tumida, ungulata, viola*
II. Test surface smooth to shiny under light microscope (microgranular, canaliculate)
  A. Surface visibly perforated [Globorotaliidae]
    1) Trochospiral coiling, non-keeled [*Globorotalia*]—species included: *anfracta, bermudezi, crassaformis, eastropacia, incisa, inflata, oscitans, palpebra, planispira, pseudopumilio, triangula, wilesi*
    2) Streptospiral coiling [*Pulleniatina*]—species included: *finalis, obliquiloculata, praecursor, primalis*
    3) Chamber flanges [*Sphaeroidinella*]—species included: *S. dehiscens, S. excavata*
  B. Surface looks imperforate [Candeinidae, Heterohelicidae]
    1) Trochospiral coiling
      a) Primary aperture with bulla
      Strongly inflated chambers [*Globigerinita*]—species included: *glutinata, iota, minuta, uvula*
      Weakly inflated subglobular or radially elongate chambers [*Turborotalita*]—species included: *humilis*
      b) Primary aperture extraumbilical [*Berggrenia*]—species included: *clarkei, praepumilio, pumilio, riedeli*
      c) Sutural apertures [*Candeina*]—species included: *C. nitida*
    2) Biserial coiling [*Streptochilus*]—species included: *S. tokelauae*

5. Order of Preference in Selecting Specimens For Illustration

I. Primary Types—Single nomenclatural type (original or replacement).
    A. Holotype—Single specimen illustrating author's original intent of a species.
    B. Lectotype—Syntypic specimen chosen later to isolate single reference specimen.
    C. Neotype—Specimen from paralectotypes chosen later to replace lost holotype.

II. Secondary Types—Specimens to support the range of variation and taxonomic concept of a given taxon.
    A. Syntypes (cotypes)—A group of specimens on which original author established his species concept, but no single specimen was designated as the final reference material.
    B. Paralectotypes—Syntypic specimen excluding the one specimen later chosen to be lectotype.

III. Tertiary Types—Other specimens originally set aside as of special taxonomic interest to supplement primary types.
    A. Paratype—Specimen(s) other than holotype on which original specific description was based.
    B. Allotype—Specimens illustrating extremes or variations deviating from designated concept.

IV. Other Types
    A. Topotype—Specimen from original locality from which a given species was described.
    B. Metatype—Topotype specimen whose identity has been confirmed by the original author.
    C. Homotype, Homeotype—Specimen compared directly by competent taxonomist with holotype, lectotype or other primary types.
    D. Ideotype—Specimen other than topotype which is identified by original author.
    E. Plesiotype—Figured specimen used subsequently to original description.
    F. Hypotype—Described or figured specimen used in publication to correct or to extend knowledge of previously described specimen.
    G. Plastotype—Replica of type specimens.

PLATES AND DESCRIPTIONS

1. 1 *Hastigerina parapelagica* SAITO and THOMPSON, 1976
(Plate 1, Figs. 1a–c, × 50; 1d, × 666)

Hastigerina pelagica (D'ORBIGNY). — BRADY, 1884 (part), p. 613, pl. 83, fig. 8 (not figs. 1–6, ?7). — BELYAEVA, 1964, p. 45, pl. 3, fig. 10, ?11–12. — SOUTAR, 1971, p. 226, pl. 13.2.1, outer two specimens.
Hastigerina rhumbleri GALLOWAY. — BÉ, 1959, p. 83, pl. 2, fig. 23. — BÉ, 1965, p. 83, text-fig. 3b, middle picture.
Hastigerinella digitata (RHUMBLER). — BANNER and BLOW, 1960b, p. 25, text-fig. 5.
Hastigerina digitata RHUMBLER. — BELYAEVA, 1964, p. 49, pl. 3, figs. 4–6.
Hastigerina parapelagica SAITO and THOMPSON, 1976, pp. 283–284, pl. 2, fig. 2; pl. 6, fig. 6.

Types: Holotype (*H. parapelagica*)—V26-128, 571 cm (Lat. 20°27′N, Long, 83°22′W, 3,868 m); holotype refigured.

Diagnosis: Test large in size, a very low trochospiral initial coil quickly becoming planispiral with growth; 5–6 subglobular chambers in the last whorl, about 10–12 chambers in all, arranged in 2–2 $1/2$ whorls. Chambers ovoid to slightly radially elongate, partially embracing with deep sutures, sutures becoming limbate near the biumbilical depressions. Aperture, a high, wide, symmetrical arch, interiomarginal, equatorial, parallel to the axis of coiling, rimmed with a wide and thick imperforate lip extending around the aperture to both the spiral and umbilical depressions. Wall calcareous, densely perforate with fine circular pores. Spines triradiate, possibly possessing twin barbs on the spine-blade edges; spines concentrated on the distal ends of the chambers; spine bases rounded-triangular.

Remarks: Although BÉ (1965) concluded that the form later called *H. parapelagica* was an intermediate growth stage of the species *H. pelagica* (with the final mature form becoming *Hastigerinopsis digitiformans*), SAITO, THOMPSON and BREGER (1976) preferred the conclusion of BANNER and BLOW (1960b, p. 26) that the three are separate species. *H. parapelagica* is larger in overall size and more lobulate than *H. pelagica*, but is less clavate than *H. digitiformans*. BELYAEVA (1964) illustrated several specimens which may be larger, more mature forms (pl. 3, figs. 4–6).

Distribution: This species is rare in marine sediments and plankton tows. It probably appeared in the very late Pleistocene or Recent, and occupies a similar tropical and deep-water habitat as *H. pelagica*.

1. 2 *Hastigerinopsis digitiformans* SAITO and THOMPSON, 1976
Plate 1, Figs. 2a–c, × 60; 2d, × 666

Hastigerina digitata (BRADY) — RHUMBLER, 1895, p. 94. — RHUMBLER, 1901, p. 70. — RHUMBLER, 1911, pp. 163–164, 202, 220, pl. 37, figs. 7–14. — SOUTAR, 1971, p. 226, pl. 13.2.1 (two middle specimens).
Hastigerinella digitata (BRADY). — CUSHMAN, 1927, p. 87, pl. 10, fig. 9 (after BRADY). — BANNER and BLOW, 1960b, pp. 24–26, text-figs. 5, 8. — BÉ, JONGEBLOED and McINTYRE, 1969, p. 1388, pl. 101, fig. 2. — HEMLEBEN, 1969, pl. 8, figs. 4–6.
Hastigerinella rhumbleri GALLOWAY, 1933, pl. 30, fig. 9 (after RHUMBLER) (*nomen nudum*). — BÉ, 1965, p. 83, text-fig. 2B, right-hand specimen. — BELYAEVA, 1964, p. 52, pl. 3, fig. 3. — JENKINS and ORR, 1972, p. 1106, pl. 37, figs. 4–6.
not *Globigerina digitata* BRADY, 1879. — BRADY, 1884 (part), pl. 80, figs. 6–10 = *Beella digitata* (BRADY).
not *Globigerina digitata* BRADY, 1879. — BRADY, 1884 (part), pl. 82, figs. 6–7 = *Globigerinella adamsi* (BANNER and BLOW).

Hastigerinopsis digitiformans SAITO and THOMPSON, 1976, pp. 284–286, pl. 2, figs. 3–4; pl. 6, fig. 5; pl. 8, fig. 2.

Types: Holotype (*H. digitiformans*)—from BANNER and BLOW, 1960b, text-fig. 8a–c. Paratype—V26-128, 571 cm (Lat. 20°27′N, Long. 83°22′W, 3,868 m) after SAITO, THOMPSON and BREGER, 1976, pl. 2, fig. 3, refigured.

Diagnosis: Test medium to very large in size, initially a low trochospire later becoming a streptospire; 6 clavate chambers per whorl; chambers initially subglobular quickly becoming radially elongate and finally elongate-clavate, frequently bifurcating and trifurcating in mature individuals. Aperture a high and wide arch, umbilical-extraumbilical, equatorial, with a thin lip. Wall calcareous, irregularly perforated with fine subcircular pores; spinose. Spines consistently triradiate in cross-section with unevenly spaced, upward-pointing two-pronged barbs along the spine-blades; spines concentrated at distal ends of chambers only, with large, upraised circular spine bases with triradiate sockets.

Remarks: SAITO, THOMPSON and BREGER (1976) discussed the validity and synonymy of this species, proposing the new genus *Hastigerinopsis* (type species *H. digitiformans*). This species can be differentiated from *H. pelagica* and *H. parapelagica* by its clavate chambers.

Distribution: This species can be considered extremely rare in marine sediments, and rare in deep plankton tows, probably because of its very fragile test and deep habitat. It probably is very late Pleistocene or Recent in appearance and similar in geographic distribution to *H. pelagica*.

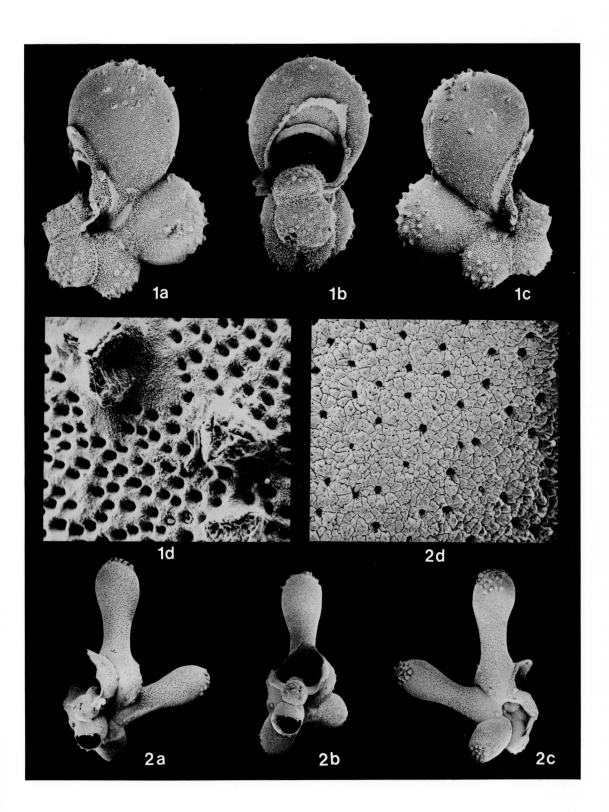

2. 1 *Hastigerina pelagica* (D'ORBIGNY), 1839
Plate 2, Figs. 1a–c, × 66; 1d, × 533

Nonionina pelagica D'ORBIGNY, 1839 (part), p. 27, pl. 3, figs. 13–14, (not figs. 1–2).
Lituola pelagica (D'ORBIGNY). — JONES and PARKER, 1860, p. 302, table 181.
Globigerina pelagica (D'ORBIGNY). — PARKER and JONES, 1865, p. 306.
Hastigerina murrayi THOMSON, 1876, p. 534, pls. 22–23. — BELYAEVA, 1964, p. 52, pl. 3, figs. 1–2.
Globigerina (Hastigerina) pelagica (D'ORBIGNY). — BÜTSCHLI, 1880, p. 202.
Hastigerina pelagica (D'ORBIGNY). — BRADY, 1884 (part), p. 613, pl. 83, figs. 1–6, 7? (not fig. 8). —
    BANNER and BLOW, 1960b, pp. 19–22, text-fig. 1 (lectotype). — JENKINS, 1971, p. 76 (emend.). —
    WALKER and VILKS, 1973, pp. 196–198, pl. 1, figs. 1–5. — SAITO and THOMPSON, 1976, pp. 282–283,
    pl. 2, fig. 1; pl. 6, fig. 4; pl. 8, fig. 9.
Hastigerina (Hastigerina) pelagica (D'ORBIGNY). — BANNER and BLOW, 1960b, *loc. cit.* (emend.).

Types: Syntype—*N. pelagica*.
Neotype—*N. pelagica* and lectotype *H. murrayi* THOMSON (1876), pl. 22, lower specimen; designated by BANNER and BLOW, 1960b, p. 21.
Plesiotype—RC8-23, trigger core top (Lat. 25°09'S, Long. 12°46'W, 3,338 m).

Diagnosis: Test large in size, initially a low trochospire becoming planispiral in the adult, about 5 subglobular chambers in each whorl. Chambers spherical to ovate, partially embracing, increasing rapidly in size as added, sutures depressed. Aperture interiomarginal, equatorial, biumbilicate, a wide and high arch often with a very thin lip. Wall calcareous, densely perforated with fine pores; spinose. Spines triradiate throughout their length with unevenly spaced upward pointing twin barbs along the spine-blade edges. Spines concentrated along equatorial periphery; spine bases large rounded triangular mounds with triradiate sockets.

Remarks: An important distinction, directly related to the phylogeny of the genus *Hastigerina,* has developed through SEM study of this species and *Globigerinella aequilateralis,* in that *pelagica* has consistently triradiate and barbed spines whereas *aequilateralis* has spines which begin round and become triradiate with growth. BOLTOVSKOY (1975) has noted that the low pore concentration and/or the organic capsule which covers the test greatly protect the species from solution, and that its fragility, rather than its susceptibility to dissolution, determines its relative rarity in sediments. Hemleben *et al.* (1979) attribute this same rarity to a structural weakening of the shell as a result of calcite resorption during gametogenesis.

Distribution: Late Miocene (N. 17) to Recent in equatorial and warm-temperate waters.

2. 2 *Globigerinella aequilateralis* (BRADY), 1839
Plate 2, Figs. 2a–c, × 66; 2d, × 533

Globigerina siphonifera D'ORBIGNY, 1839a, p. 83, pl. 4, figs. 15–18.
cf. *Globigerina hirsuta* D'ORBIGNY, 1839b, p. 133, pl. 2, figs. 4–6.
Cassidulina globulosa EGGER, 1857 (part), p. 296, pl. 11, fig. 4 (only).
Globigerina aequilateralis BRADY, 1879, p. 285. — BRADY, 1884, p. 605, pl. 80, figs. 18–21.
Globigerina aequilateralis (BRADY) var. *involuta* CUSHMAN, 1917, p. 662. Plates published 1921, p. 293, text-figs. 11a–c.
Globigerinella aequilateralis (BRADY). — CUSHMAN, 1927, p. 87. — BRADSHAW, 1959, p. 38, pl. 7, figs.
    1–2. — WALKER and VILKS, 1973, pp. 196–198, pl. 1, figs. 6–7. — SAITO, THOMPSON and BREGER,

Plate 2 *Hastigerina pelagica* (D'ORBIGNY), 1839
Globigerinella aequilateralis (BRADY), 1839

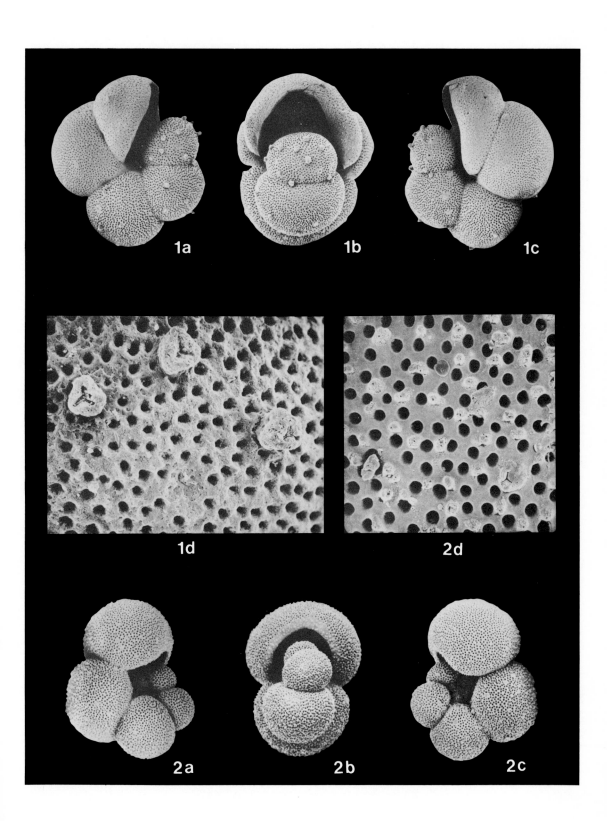

1976, pp. 281–282, pl. 3, figs. 1–2; pl. 6, fig. 7; pl. 8, figs. 3, 8.
Globigerina (Globigerinella) aequilateralis (BRADY).   —   COLOM, 1952, pp. 10, 42, table, pl. 8, figs. 6–7.
Globigerina (Globigerinella) involula CUSHMAN.   —   HOFKER, 1956, p. 224, pl. 33, figs. 33–34.
Hastigerina aequilateralis (BRADY).   —   BOLLI, LOEBLICH and TAPPAN, 1957, p. 29, pl. 3, fig. 4.
Hastigerina aequilateralis involuta (CUSHMAN).   —   BLOW, 1959, p. 171, pl. 8, fig. 32.
Hastigerina (Hastigerina) siphonifera (D'ORBIGNY).   —   BANNER and BLOW, 1960b, pp. 22–23, text-figs.
2a–c, 32b (lectotype) (*nomen oblitum*).
Globigerinella involuta (CUSHMAN).   —   BERMÚDEZ, 1961, p. 1212, pl. 6, fig. 11; pl. 7, fig. 1.
Globigerinella siphonifera (D'ORBIGNY).   —   PARKER, 1962, p. 228, pl. 2, figs. 22–28.
Hastigerina involuta (CUSHMAN).   —   BERMÚDEZ and BOLLI, 1969, pp. 181–182, pl. 17, figs. 6–7.

Types: Syntype and lectotype (*siphonifera*)—from D'ORBIGNY's collection from Cuba.

Lectotype (*aequilateralis*)—BRADY, 1884, pl. 80, fig. 19, Challenger Station 224.

Plesiotype—V28-222, trigger core top (Lat. 11°19'S, Long. 174°32'E, 2,933 m).

Diagnosis: Test medium to large in size, very low trochospire becoming planispiral, about 5–6 globular chambers per whorl. Chambers spherical to ovate, slightly embracing, increasing rapidly in size as added, chambers of later whorls often covering earlier whorls, sutures distinct and depressed. Aperture interiomarginal, biumbilical, a low but wide symmetrical arch, with no apparent rim or lip. Wall calcareous, densely perforated with large circular pores, spinose. Spines have round cross-sections at the base becoming triradiate with growth, spine bases upraised circular mound with circular socket.

Remarks: This species can be observed to have a raised symmetrical to highly assymmetrical coiling depending on the height of the trochospire. Juvenile forms are quite trochospiral and are often referred to as *Globorotalia obesa* BOLLI, the ancestral form of *G. aequilateralis*. Small, uncompleted chambers often occur in the final whorl and give the test the appearance of unraveling. Under the microscope, *G. aequilateralis* is more hispid than *H. pelagica,* and more tightly coiled and planispiral than *G. calida.* CUSHMAN (1917) separated very tightly coiled forms as the variety *involuta.*

Distribution: BLOW (1969) gives the range of this species (listed as   *H. (H.) siphonifera siphonifera*) as late Middle Miocene to Recent. It is a tropical to subtropical form.

3. 1 *Globigerinella adamsi* (BANNER and BLOW), 1959
Plate 3, Figs. 1a–c, × 66; 1d, × 533

Globigerina digitata BRADY, 1884 (part), pl. 82, figs. 6–7 (not pl. 80, figs. 6–10).
Hastigerinella digitata (BRADY). — CUSHMAN, TODD and POST, 1954, p. 369, pl. 91, figs. 9–10.
Globigerinella sp., BRADSHAW, 1959, p. 38, pl. 7, figs. 3–4.
Hastigerina (Bolliella) adamsi BANNER and BLOW, 1959, pp. 13–14, figs. 4a–d. — BANNER and BLOW, 1960b, p. 24, text-figs. 4a–c.
Globigerinella adamsi (BANNER and BLOW). — PARKER, 1962, pp. 227–228, pl. 2, figs. 19–21. — BÉ, JONGEBLOED and MCINTYRE, 1959, p. 1391, pl. 161, fig. 3. — SAITO, THOMPSON and BREGER, 1976, p. 281, pl. 1, fig. 3; pl. 6, fig. 3; pl. 7, fig. 7.
Bolliella adamsi BANNER and BLOW — LOEBLICH and TAPPAN, 1964, p. C665, text-fig. 531, figs. 9a–c.
Hastigerina digitata RHUMBLER. — BELYAEVA, 1964, p. 49, pl. 3, figs. 7–9.
Beella discors McCULLOCH, 1979, p. 422, pl. 174, figs. 14–16.

Types: Holotype—Challenger Station 191 A (Lat. 05°26′S, Long. 133°19′E, 580 fm).
Plesiotype—V28-222, trigger core top (Lat. 11°19′S, Long. 174°32′E, 2,933m).

Diagnosis: Test large to very large in size, a very low trochospire with about 6 radially elongate chambers per whorl. Chambers initially spherical, soon becoming ovate and then quite radially elongate tapering to a pointed distal end which frequently has one or two small knob-like projections. Chambers increase rapidly in size as added and later whorls do not usually touch earlier whorls, sutures depressed. Aperture interiomarginal, equatorial, a high and wide symmetrical arch with a distinct lip. Wall calcareous, densely perforated with large circular pores, spinose. Spines round-becoming-triradiate, emerging from rounded-upraised spine bases.

Remarks: The digitate chambers and low trochospiral coiling of this form make it quite distinctive from *H. digitiformans, G. aequilateralis* or *B. digitata.* Juvenile forms are very similar to *G. calida* to which *G. adamsi* is undoubtedly related and may even be conspecific.

Distribution: Late Pleistocene (N. 23) to Recent, Indo-Pacific equatorial waters only.

3. 2 *Beella digitata* (BRADY),1879
Plate 3, Figs. 2a–c, × 66; 2d, × 533

Globigerina digitata BRADY, 1879 (part), p. 286. — BRADY, 1884 (part), p. 599, pl. 80, figs. 6–10 (not pl. 82, figs. 6–7). — PHLEGER, PARKER and PEIRSON, 1953, p. 12, pl. 1, figs. 9–10. — PARKER, 1962, p. 222, pl. 1, figs. 20–25. — JENKINS and ORR, 1972, p. 1087, pl. 6, figs. 7–8.
Globigerina cf. *digitata* BRADY. — CUSHMAN and STAINFORTH, 1945, p. 6, 68, pl. 13, fig. 5.
Globigerinella digitata (BRADY). — SCHOTT, 1937, p. 59.
Globigerina (Globigerinella) digitata BRADY. — HOFKER, 1956 (part), p. 225, pl. 34, figs. 6–7 (only).
Hastigerinella digitata (BRADY). — GALLOWAY, 1933, p. 333. — BOLLI, LOEBLICH and TAPPAN, 1957, p. 32, pl. 5, figs. 3a–g. — RÖGL and BOLLI, 1973, p. 567, pl. 4, figs. 16–17; pl. 13, figs. 6–9.
Globorotalia (Hastigerinella) digitata (BRADY). — BANNER and BLOW, 1959, p. 16, text-fig. 4e (lectotype).
Globorotalia (Beella) digitata (BRADY). — BANNER and BLOW, 1960b, pp. 26–27, text-fig. 11.
Beella digitata (BRADY). — LOEBLICH and TAPPAN, 1964, pp. C669–670, text-fig. 537. — SAITO, THOMPSON and BREGER, 1976, pp. 280–281, pl. 1, fig. 1; pl. 6, fig. 1.
Beella chathamensis McCULLOCH, 1979, p. 422, pl. 174, figs. 9, 12, 13.
Beella guadalupensis McCULLOCH, 1979, pp. 422, 423, pl. 174, fig. 11.
Hastigerinella (?) *frailensis* McCULLOCH, 1979, p. 423, pl. 174, fig. 10.

Types: Type locality not given, figured specimens from Challenger Stations 276, 338, 191A.

         Syntype—Challenger Station 191A.

         Lectotype—Challenger Station 338, North Atlantic, BRADY (1884), pl. 80, fig. 10.

         Plesiotype—V28-222, trigger core top (Lat. 11° 19′S, Long. 174° 32′E, 2,933 m).

Diagnosis: Test size highly variable, medium-height trochospire becoming streptospire, usually about 4 chambers in the final whorl, 10–15 globular to radially elongate chambers in all, arranged in about $2-2^1/_2$ whorls. Chambers initially subspherical soon becoming ovate and in later whorls radially elongate, increasing rapidly in size as added, partially embracing, sutures distinct, depressed. Aperture interiomarginal, umbilical-extraumbilical, usually with a wide and high rectangular arch surrounded with a prominent imperforate lip in adults; some additional apertural openings may form due to irregular attachment of last chamber to the earlier whorl, but are not typical secondary supplementary apertures. Wall calcareous, densely perforated with irregularly spaced, usually multiple, clusters of 2–6 porelets separated by narrow walls, possibly spinose. Spines observed were round becoming triradiate.

Remarks: This form is quite variable in morphology due to the chamber shape which varies through growth and the streptospiral coiling mode. Small specimens have ovoid chambers and a very low trochospire while adult forms have quite radially elongate chambers and a medium height streptospire. The most distinct feature of the test is its unusual multiple-porelet clusters, otherwise seen only on *Neogloboquadrina pachyderma*.

Distribution: Evolved from *Globigerina praedigitata* PARKER in the Late Pliocene (N. 21) and continues to Recent. PARKER (1962) reports it as far south in the Pacific as 44° S, the lower latitude forms being lower-spired than those from higher latitudes. It is seldom an abundant form, but persists in small numbers throughout its biogeographic range.

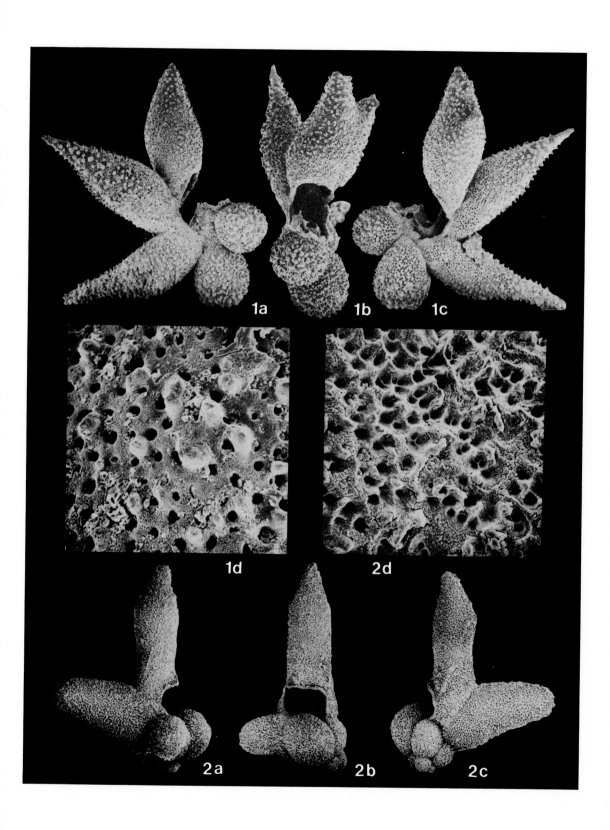

4. 1 *Globigerina bermudezi* SEIGLIE, 1963
Plate 4, Figs. 1a–c, × 80; 1d, × 800

Globigerina bermudezi SEIGLIE, 1963, pp. 90–91, pl. 1, figs. 1–8. — BERMÚDEZ and BOLLI, 1969, pp. 149–150, pl. 1, figs. 4–6. — RÖGL and BOLLI, 1973, p. 562, pl. 2, figs. 11–19; pl. 12, fig. 4.

Types: Holotype—locality G63-01, Station 22 (Lat. 10°31'40"N, Long. 64°18'25"W, 290 m).

Plesiotype—RC8-39, 30 cm (Lat. 42°53'S, Long. 42°21'E, 4,330 m).

Diagnosis: Test large, a medium to very high trochospire, 5 chambers in the final whorl, about 12–16 globular chambers arranged in about $2^1/_2$–3 whorls. Chambers spherical except for the final and often penultimate chambers which become very inflated ventrally and extend over the umbilicus, sutures distinct, deep. Aperture umbilical-extraumbilical, a large and variable arch whose shape is controlled by the ventral extent of the last chamber; secondary openings often form by irregular contact of the last chamber with earlier chambers but are not considered supplementary apertures. Walls calcareous, unevenly perforated with circular pores, spinose. Spines round to round-becoming-triradiate emerging from large, upraised spine bases.

Remarks: Due to the range of height of coiling and bizarre morphology of the final chambers, this species is highly variable in typical morphology. In this aspect, however, it can be distinguished from other *Globigerina*.

Distribution: Late Pleistocene to Recent, living in the Cariaco Basin (RÖGL and BOLLI, 1973), and found in the Caribbean (SEIGLIE, 1963). The authors have noted very rare occurrences in the south central Pacific.

4. 2 *Globigerinella calida* (PARKER), 1962
Plate 4, Figs. 2a–c, × 80; 2d, × 800

Globigerina sp. BRADSHAW, 1959, p. 38, pl. 6, figs. 19, 26–28.
Globigerina calida PARKER, 1962, p. 221, pl. 1, figs. 9–13, 15. — PARKER, 1967, p. 149, pl. 18, fig. 11 (not fig. 6–10, 12).
Globigerina calida PARKER subsp. *calida* PARKER. — BLOW, 1969, p. 380, pl. 13, figs. 9, 10 (emend.).
Globigerinella calida (PARKER). — SAITO, THOMPSON and BREGER, 1976, p. 282, pl. 1, fig. 2; pl. 6, fig. 2; pl. 8, fig. 1.

Types: Holotype—Downwind BG130, 0–4 cm (Lat. 14°44'S, Long. 112°06'W, 3,120 m).

Plesiotype—V18-272, trigger core top (Lat. 12°47'S, Long. 161°03.5'W, 3,160 m).

Diagnosis: Test size variable, a low trochospire, 4–6 subglobular to slightly radially elongate chambers in the final whorl, about 12–15 chambers in all arranged in about $2^1/_2$ whorls. Chambers initially spherical becoming sub-spherical, size increasing moderately as chambers are added, chambers loosely embracing so that in mature specimens the final chamber is almost completely detached from the previous whorl; sutures deep. Aperture umbilical-extraumbilical, a low asymmetrical arch with a thin lip. Wall calcareous, densely perforated with large circular pores; spinose. Spines round to round-becoming-triradiate, emerging from round upraised spine bases.

Remarks: Adult forms of this species are readily identified by the very loose coiling

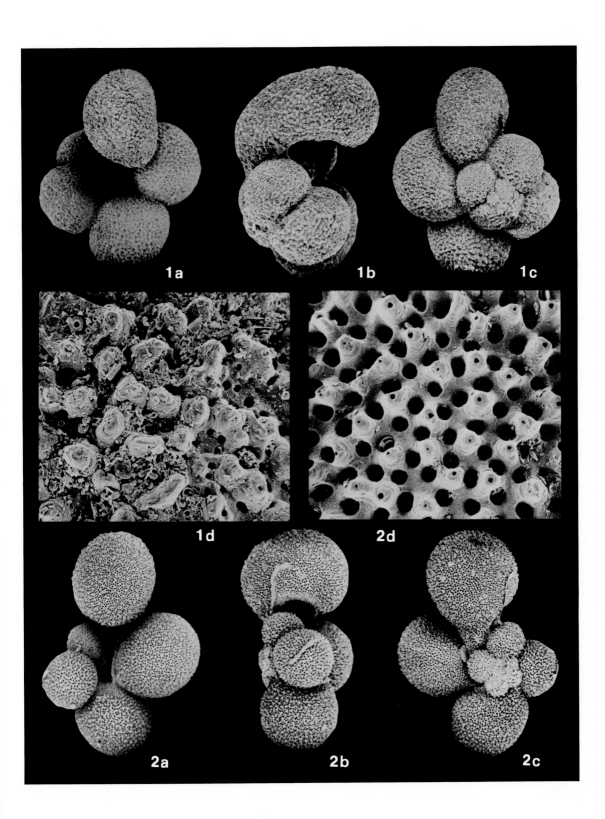

1a 1b 1c

1d 2d

2a 2b 2c

resulting in almost complete separation of the final chamber and the previous whorl. In this respect, it resembles, and may be related to, *G. adamsi,* but lacks digitate chambers. Juvenile *G. calida* are very difficult to distinguish from juvenile *G. bulloides* except for the spine morphology which is visible under the SEM: *G. calida* has round-becoming-triradiate spines whereas *G. bulloides* has simple round spines.

Distribution: Late Pleistocene (N. 73) to Recent. PARKER (1962) reports living forms north of 30–35° S latitude and BRADSHAW reports it to 40° N latitude in the Pacific, and RöGL and BOLLI (1973) found it in the eastern Caribbean.

5. 1 *Globigerina cariacoensis* Rögl and Bolli, 1973
Plate 5, Figs. 1a–c, × 133; 1d, × 1333

Globigerina megastoma cariacoensis Rögl and Bolli, 1973, p. 564, pl. 2, figs. 1–10; pl. 12, figs. 5–6; text-figs. 4a–c.
Globigerina cariacoensis Rögl and Bolli. — Poore and Berggren, 1975, p. 292, pl. 5, figs. 9–12.

Types: Holotype—*G. cariacoensis* DSDP Site 147, Core 2, Section 1, top (Lat. 10° 42'48''N, Long. 65° 10'48''W).

Plesiotype—*G. cariacoensis* V12-98, 0–10 cm (Lat. 10° 47.1'N, Long. 65° 06.8'W, 736 m).

Diagnosis: Test large, a high trochospire, 4 globular chambers in the final whorl, about 11–17 chambers arranged in about 3 whorls. Chamber shape spherical to ovoid with a tendency to become radially elongate ventrally in the final whorl, sutures deeply incised, chambers increasing slowly in size as added. Aperture umbilical, open, a low arch with no modifications. Wall calcareous, coarsely perforated; spines simple, rounded with circular cross-section and raised spine bases.

Remarks: This species differs from *G. megastoma* by having a wider aperture, more elongate chambers and tropical ecology.

Distribution: Late Pleistocene to Recent, tropical Atlantic (only reported occurrences).

5. 2 *Globigerina megastoma* Earland, 1934
Plate 5, Figs. 2a–c, × 133; 2d, × 1333

Globigerina megastoma Earland, 1934, pp. 177–178, pl. 8, figs. 9–12. — Banner and Blow, 1960a, pp. 14–15, pl. 1, figs. 3a–c (lectotype).
?*Globigerina partidiana* McCulloch, 1979, p. 415, pl. 173, fig. 3.

Types: Syntypes—Discovery Station 385 (Lat. 58 41'S, Long. 64°43'30''W, 3,638 m); William Scoresby Stations, WS 204 (Lat. 54°27'S, Long, 54°22'W, 3,388 m); WS 403 (Lat. 59°40'S, Long. 54 35'W, 3,721 m); WS 469 (Lat. 56°42'S, Long. 57°00'W, 3,959 m).

Lectotype—Original materials, Earland's fig. 9, from Station 385.

Plesioytpe—V19-29, 300 cm (Lat. 03°35'S, Long. 83°56'W, 3,157 m).

Diagnosis: Test medium to large sized, medium to high trochospire, about 5 globular chambers in the final whorl, about 11 chambers arranged in about 2 1/2 whorls. Chamber well separated, sutures deeply incised, chambers increasing slowly in size, with a tendency for radial elongation on the ventral side of the final whorl. Aperture interiomarginal, umbilical, open, occasionally with an imperforate lip. Wall calcareous, finely perforated, spinose. Spines simple, rounded in circular cross-section raised spine bases.

Remarks: This form is encountered only rarely in temperate assemblages and can usually be recognized by its distinctive ovoid chambers and fairly high trochospire.

Distribution: Late Pliocene (N. 21) to Recent, temperate to subpolar waters of the southern hemisphere.

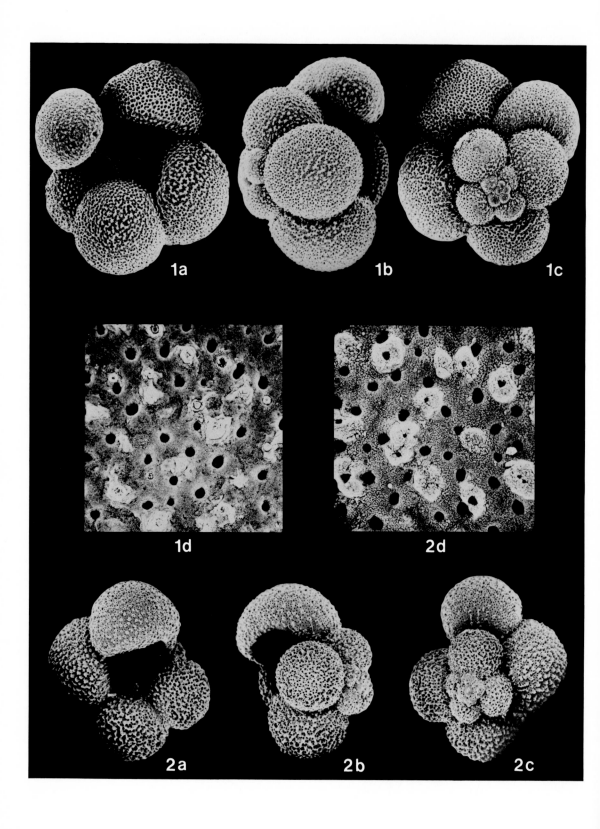

6. 1 *Globigerina umbilicata* ORR and ZAITZEFF, 1971
Plate 6, Figs. 1, 2a–b, × 120; 2c, × 1200

Globigerina diplostoma BERMÚDEZ and BOLLI, 1969 (not REUSS, 1850), p. 151, pl. 2, figs. 1–3.
Globigerina umbilicata ORR and ZAITZEFF, 1971, pp. 17–19, pl. 1, figs. 1–3.
Globigerina bulloides umbilicata ORR and ZAITZEFF. — RÖGL and BOLLI, 1973, p. 563, pl. 1, figs. 19–20; pl. 11, fig. 11.

Types: Holotype (*G. umbilicata*)—from "a thick grey mudstone 2100 feet south of Fleener Creek on Centerville Beach in SW 1/4, NE 1/4,SW 1/4, Section 13, T2N, R3W, Del Rio Formation (Wildcat Group), Eureka, Humboldt County, California."

Plesiotype (*G. umbilicata*)—RC11-187, trigger core top (Lat. 47°08.7′N, Long. 130°06.7′W, 2,670 m).

Diagnosis: Test medium to large in size, 4–6 but usually 5 globular chambers in the final whorl, about 10–12 chambers in all arranged in a low trochospire of about $2\,{}^1/_2$ whorls. Chambers spherical to subspherical, increasing slowly in size as added, well separated giving a slight lobulate equatorial peripheral outline, sutures distinct, well incised; the final chamber is often reduced in size or incompletely inflated. Aperture umbilical, a large and open arch with a thin lip. Wall calcareous, perforated; spines have circular cross-section.

Remarks: This species can be distinguished from *G. bulloides* by its greater number of chambers in the final whorl and well-depressed spiral suture.

Distribution: Originally reported as Pliocene only; specimens of RÖGL and BOLLI and those here are Pleistocene; habitat the same as *G. bulloides*.

6. 2 *Globigerina exumbilicata* HERMAN, 1974
Plate 6, Figs. 3a–c, × 200; 3d, × 2000

Globigerina exumbilicata HERMAN, 1974, pp. 299–300, pl. 17, fig. 1; pl. 18, figs. 1–5; pl. 19, figs. 1–5.

Types: Holotype—from Drift Station Alpha II, 144 cm (Lat.83°52′N, Long.168°12′W, 1,521 m) (Location given as Drift Station A 2 II)

Topotype—from Drift Station A II, 144–145 cm.

Diagnosis: Test small, a well-lobulated trochospire of 10–12 globular chambers arranged in about $2\,{}^1/_2$ whorls, $4\,{}^1/_2$–5 chambers in the final whorl, chambers well separated, sutures radial and well incised. Aperture umbilical, a subquadrate opening partially obscured by the ventral inflation of the base of the final chamber which has a thin lip. Wall calcareous, very finely perforated, densely hispid with upraised spine bases, spines not observed.

Remarks: In earlier papers, HERMAN (1974 refs.) included this species with *G. quinqueloba,* although it may be distinguished by the more lobulate test and absence of the bulla-like extension of the base of the final chamber that often obscures the aperture.
bulla-like extension of the base of the final chamber that often obscures the aperture.
In her 1980 paper, HERMAN concluded this taxon to be a junior synonym of *Globigerina quinqueloba egelida* CIFELLI and SMITH. This species is included here pending further

taxonomic study by other workers, but the present authors are inclined to agree with HERMAN's conclusion of 1980.

Distribution: HERMAN has noted this species only from Quaternary sediments of the Arctic Ocean. It appears to be quite susceptible to dissolution.

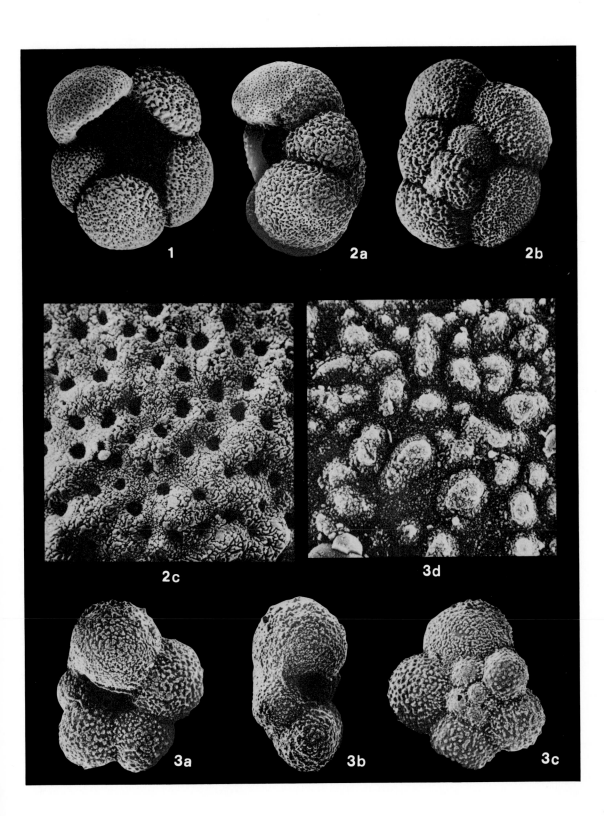

7. 1 *Globigerina bulloides* D'ORBIGNY, 1826
Plate 7, Figs. 1a–c, × 133; 1d, × 1333

"*Polym. Tuberosa et Globuliferae*" SOLDANI, 1791 (part), p. 117, pl. 23, fig. O (not figs. H, I, and P *fide* BANNER and BLOW, 1960a, p. 3).
Globigerina bulloides D'ORBIGNY, 1826, p. 277 (no figures), list no. 1. — D'ORBIGNY, 1839b, p. 132, pl. 2, figs. 1–3 (not fig. 28). — PARKER, JONES and BRADY, 1871, pl. 11, fig. 112 (after SOLDANI, 1791). — BANNER and BLOW, 1960a, pp. 3–4, pl. 1, figs. 1a–c (lectotype), 4.
Globigerina diplostoma REUSS, 1850, p. 373, pl. 47, figs. 9, 10; pl. 48, fig. 1?
Pylodexia bulloides (D'ORBIGNY). — EHRENBERG, 1872, p. 276 (synonymy only).
Globigerina quadrilatera GALLOWAY and WISSLER, 1927, p. 44, pl. 7, fig. 11.

Types: Syntypes (*G. bulloides*)—from the Adriatic near Rimini (Lat. 44°05'N, Long. 12°06'E), but may have been reworked from older horizons.
Lectotype—from remains of D'ORBIGNY's materials.
Plesiotype—V10-12, trigger core top (Lat. 40°48'N, Long. 12°45.3'E, 1,957 m).

Diagnosis: Test size variable, low to medium height trochospire, well lobulated, 3–5 globular chambers in the final whorl, about 8–10 chambers in all arranged in 2–2 $1/2$ whorls, chamber shape spherical to slightly ovoid, chambers well separated with deep sutures, chamber size increases slowly. Aperture umbilical, interiomarginal, a high, symmetrical arch occasionally with a thin rim-like lip. Wall calcareous, uniformly and densely perforated; spines simple and with circular cross-sections.

Remarks: BANNER and BLOW (1960a) provided a thorough study of the synonymy of this taxon. BANDY (1972a) called attention to the latitudinal variation in the number of chambers in the final whorl often giving the idea of separate subspecies. Typical specimens have an average of 4 chambers in the final whorl, but the rate of chamber growth leads to a variety of peripheral lobulation types. It can be distinguished from *G. falconensis* by its lack of the prominent lip and from *G. calida* by its simple circular spines. Our examination of the holotype of *G. quadrilatera* shows it to be a four-chambered variety of *G. bulloides* with a slightly reduced final chamber; it should be noted that the figure by GALLOWAY and WISSLER suggests that the chambers of the final whorl are distally blunted, although the specimen has spherical chambers.

Distribution: Late Miocene (N. 16) to Recent, subtropical to polar waters.

7. 2 *Globigerina falconensis* BLOW, 1959
Plate 7, Figs. 2a–c, × 133; 2d, × 1333

Globigerina falconensis BLOW, 1959, p. 177, pl. 9, figs. 40a–c, 41. — BRÖNNIMANN and RESIG, 1971, pp. 1295–1296 (comments on holotype).

Types: Holotype (*G. falconensis*)—BLOW's sample RM19285, auger line near Pozón, Eastern Falcón, Venezuela (Miocene).
Plesiotype—Caryn 22-6 trigger core top (Lat. 32°16'N, Long. 64°38. 75'W, 1,509 m).

Diagnosis: Test small to medium in size, a low or medium height trochospire with 3–5 globular chambers in the final whorl, about 10–16 chambers in all arranged in about 2 $1/2$ whorls. Chamber shape spherical, the last chamber typically smaller than the pre-

Plate 7 *Globigerina bulloides* D'ORBIGNY, 1826
Globigerina falconensis BLOW, 1959

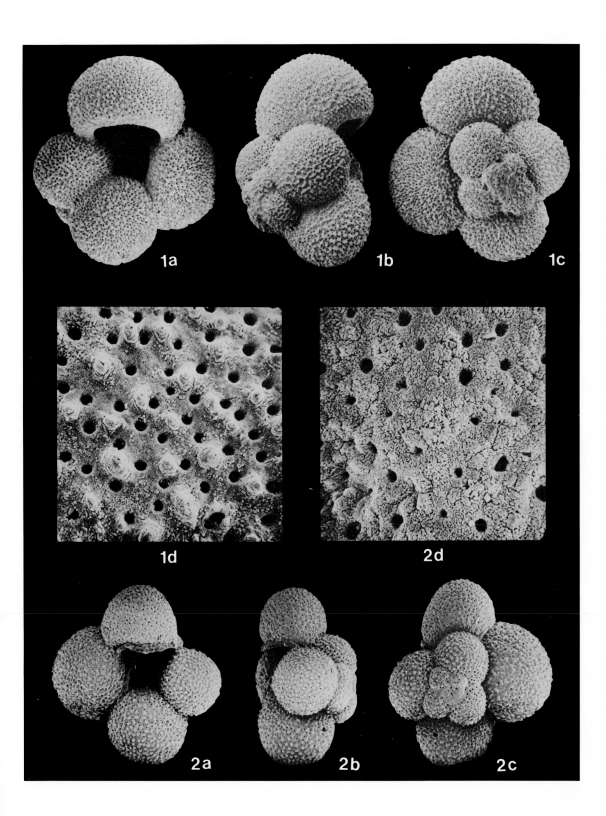

ceding one and ovoid in shape, chambers well separated by deeply incised sutures, chamber size increases slowly as added. Aperture umbilical, interiomarginal, partially to completely obscured by the base of the final chamber with its thick imperforate lip. Wall calcareous, coarsely perforated; spines simple and have circular cross-sections.

Remarks: This species is easily identified by the prominent last chamber configuration with its thick lip. For further discussion, see MALMGREN and KENNETT (1977).

Distribution: Early Miocene (N. 7) to Recent, subtropical to temperate.

8. 1 *Globorotalia incisa* BRÖNNIMANN and RESIG, 1971
Plate 8, Figs. 1a–c, × 180; 1d, × 1500

Globorotalia (Turborotalia) incisa BRÖNNIMANN and RESIG, 1971, pp. 1278–1279, pl. 45, figs. 5, 7; pl. 46, figs. 1–8.

Globorotalia (Turborotalia) pseudopachyderma BRÖNNIMANN and RESIG, 1971 (not *Globorotalia (Turborotalia) pseudopachyderma* CITA, PREMOLI-SILVA, and ROSSI, 1965), pl. 34, fig. 6; pl. 45, fig. 8.

Globigerina fossulata POAG, 1971, p. 257, figs. 8–12e.

Globigerina incisa (BRÖNNIMANN and RESIG). — POAG and VALENTINE, 1976, p. 200, pl. 11, figs. 1–9.

Types: Holotype—DSDP Leg 7, Hole 62.1, Core 8, Section 3, 15–17 cm.

Plesiotype—V28-239, 1,494 cm (Lat. 03° 15′N, Long. 159° 11′N, 3,490 m).

Diagnosis: Test small to medium sized, a low trochospire, 3–4 globular chambers in the final whorl, about 12 chambers in all arranged in $2\,^1/_2$ whorls. Chambers spherical to subspherical, closely packed, increasing moderately in size giving the test a quadrate outline, sutures very deep and incised. Aperture umbilical-extraumbilical, a very low arch bordered by a thick rim-like lip. Wall calcareous, irregularly perforated, granular.

Remarks: Although BRÖNNIMANN and RESIG (1971) related their species *G. incisa* to *G. pseudopachyderma,* the specimens of the latter chosen by them also seem to be *G. incisa.* From their diagnosis, the identification of *G. incisa* using its prominent, deeply incised sutures and strong apertural lip seem to be adequate criteria. POAG and VALENTINE (1976) believe this species to be spinose.

Distribution: Late Miocene (N. 18) to ? Pliocene (N. 20), observed in the western equatorial Pacific only.

8. 2 *Globorotalia palpebra* BRÖNNIMANN and RESIG, 1971
Plate 8, Figs. 2a, 3a–b, × 180; 2b, × 1500

Globorotalia (Turborotalia) palpebra BRÖNNIMANN and RESIG, 1971, p. 1280, pl. 3, fig. 3.

Types: Holotype—DSDP Leg 7, Hole 62.1, Core 6, Section 5, 15–17 cm (Lat. 01° 52.2′N, Long. 141° 56.3′E, 2,607 m).

Plesiotype—DSDP Hole 289, Core 4, core catcher (Lat. 00° 29.92′S, Long. 158° 30.69′E, 2,206 m).

Diagnosis: Test small, a low trochospire, 4 globular chambers in the final whorl, 8–10 chambers in all arranged in about 2–$2\,^1/_2$ whorls. Chambers spherical to slightly radially elongate, sutures well depressed. Aperture umbilical-extraumbilical, the apertural face of the last chamber bends upward to form a large apertural lip. Wall calcareous, densely perforated, pustulate.

Remarks: BRÖNNIMANN and RESIG (1971) used the pustulate surface and globorotaliid aperture to classify this species and separate it from the "polygonal surface" and umbilical aperture of *G. falconensis.* On the basis of one view of one specimen, the morphologic variation of this species is difficult to assess and our choice of a plesiotype may not be accurate.

Distribution: BRÖNNIMANN and RESIG (1971) record this species from Late Miocene (N. 17) to Late Pleistocene (N. 22) in DSDP Hole 62. 1.

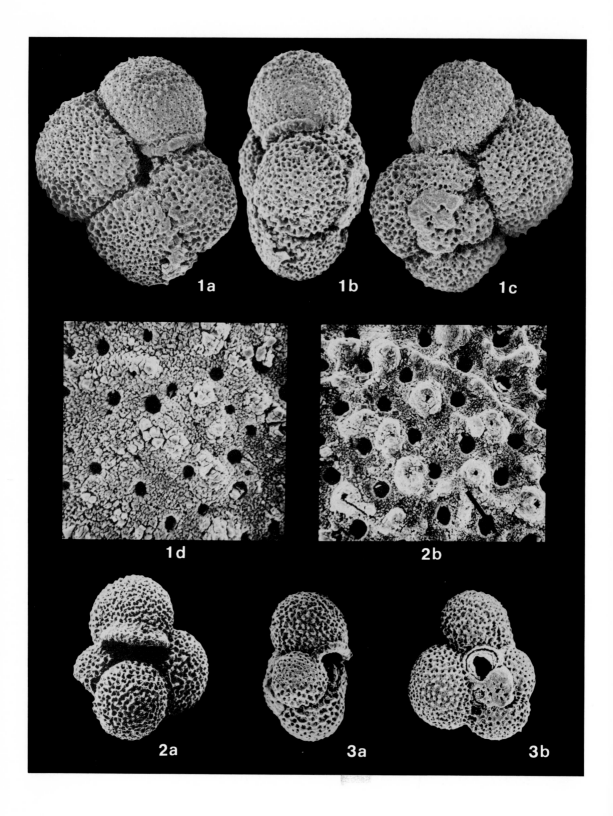

9. 1 *Globigerina atlantisae* Cifelli and Smith, 1970
Plate 9, Figs. 1a–c, × 214; 1d, × 1785

Globigerina radians Egger. — Parker, 1958, p. 278, pl. 5, fig. 10.
Globigerina atlantisae Cifelli and Smith, 1970, pp. 17–18, pl. 1, figs. 1–3.
Turborotalita atlantisae Cifelli and Smith. — Iaccarino and Salvatorini, 1979, pl. 7, figs. 25–26.

Types: Holotype—Station 18, Atlantis II Cruise 13 (Lat. 41°39'N, Long. 50°00W, 252 m, plankton tow).

Topotype—V23-15, trigger core top (Lat. 43°29'N, Long. 45°15'W, 4,415 m).

Diagnosis: Test small in size, fairly lobulate, about 4–5 globular chambers in the final whorl, about 8–14 chambers in all arranged in about 2–3 trochospiral whorls. Chambers spherical or slightly flattened radially, increasing rapidly in size as added, final chamber slightly overhanging the umbilicus; sutures incised and distinct. Aperture interiomarginal, umbilical-extraumbilical, a low arch which extends almost to the periphery, bordered by a prominent flap-like lip on the final chamber. Wall calcareous, finely perforated, thin finely hispid and possibly spinose.

Remarks: Cifelli (1965, p. 17) synonymized Parker's (1958) *G. radians* with *G. atlantisae,* although noting considerable similarity of the illustrations with *G. iota. G. atlantisae* most clearly resembles their *G. egelida,* although differing, according to Cifelli and Smith, in its greater spiral suture curvature on the umbilical side, slightly more radially flattened chambers and a prominent tendency for an "indentation" to develop in the periphery below and adjacent to the final chamber due to different rates of chamber size increase.

Distribution: Noted only in plankton tows, and, if the seabed specimen shown here is conspecific, from Recent sediments of the northwest Atlantic.

9. 2 *Globigerina egelida* Cifelli and Smith, 1970
Plate 9, Figs. 2a–c, × 214; 2d, × 1785

Globigerina cf. *quinqueloba* Natland. — Todd and Brönnimann, 1957, p. 40, pl. 12, figs. 2, 3. — Bé, 1959, p. 83, pl. 1, figs. 21–22.
Globigerina aff. *quinqueloba* Natland. — Cifelli, 1965. p. 13, pl. 2, figs. 3–4.
Globigerina quinqueloba Natland subsp. *egelida* Cifelli and Smith, 1970, pp. 32–34, pl. 3, figs. 4–7.

Types: Holotype—Station 26, Atlantis II Cruise 13 (Lat. 45°29'N, Long. 43°00'W, 251 m, plankton two).

Topotype—V27-13, trigger core top (Lat. 42°02.8'N, Long. 51°05'W, 3,294 m).

Diagnosis: Test small in size, $4^{1}/_{2}$–5 globular chambers in the final lobulate whorl, about 14 chambers in all, arranged in up to 4 low trochospiral whorls. Chambers sub-spherical, enlarging fairly rapidly in size as added, the final chamber overhangs the umbilicus slightly; sutures distinct, depressed. Aperture interiomarginal, umbilical to umbilical-extraumbilical bordered by a thin rim-like lip at the base of the final chamber. Wall calcareous, thin, finely perforated and possibly spinose.

Remarks: Cifelli and Smith (1970) distinguished *G. egelida* from the holotype of *G. quinqueloba* by the latter being more thick walled, compact and possessing less incised sutures and less lobulate periphery.

Distribution: Noted only in plankton tows and if this specimen is conspecific, from Recent sediments of the northwest Atlantic.

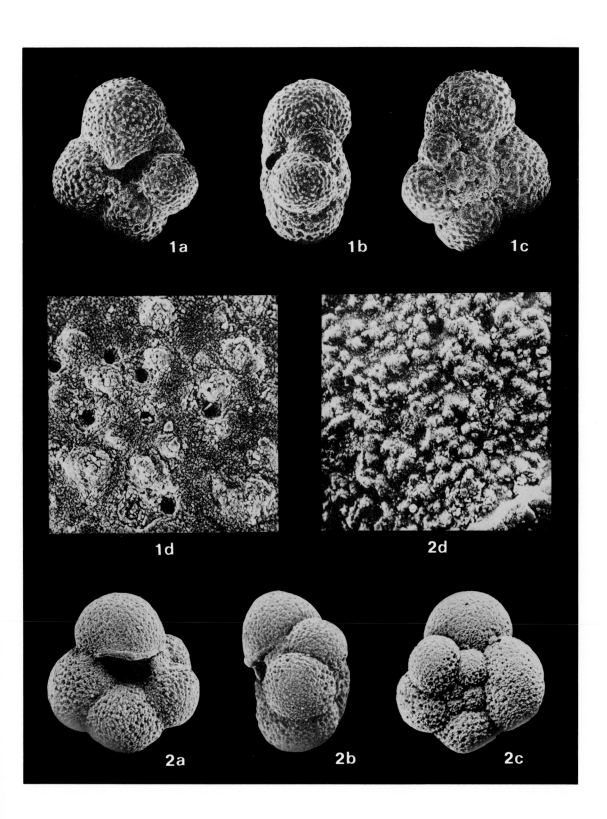

10 *Globigerina quinqueloba* NATLAND, 1938
Plate 10, Figs. 1a–c, 2a–c, × 214; 1d, 2d, × 1785

Globigerina quinqueloba NATLAND, 1938, p. 149, p. 6, figs. 7a–c. — PARKER, 1962, p. 225, pl. 2, fig. 7–16.
— HERB, 1968, p. 476, pl. 3, figs. 3a–c. — HEMLEBEN, 1969, pl. 17, fig. 4. — RÖGL and BOLLI,
1973, p. 564, pl. 4, figs. 10–12.
Globigerina groenlandica STSCHEDRINA, 1946, p. 145 (English p. 148), pl. 4, figs. 23a–b.

Types: Holotype (*quinqueloba*)—from NATLAND's Locality 118 (Lat. 33°22′20″N, Long. 118°18′31″W, 152 m).

Holotype (*groenlandica*)—from STSCHEDRINA's Station 4 (1935), (Lat. 76° 21′N, Long. 358°37′E, 3,000 m).

Plesiotype—V22-105, bottom of core (Lat. 47°00′S, Long. 13°23′W, 2,067 m).

Diagnosis: Test small in size, 5 globular chambers in the final whorl, about 15 chambers in all arranged in about 2 $^1/_2$ low trochospiral whorls. Chambers subglobular, slightly flattened radially, final chamber of mature specimens typically partially or completely covers umbilicus like a bulla, rarely with one or more infralaminal apertures; sutures distinct, slightly incised. Aperture umbilical-extraumbilical, interiomarginal, frequently obscured by final chamber with its rim-like basal lip. Wall calcareous, thick in mature specimens, finely perforated, hispid and spinose.

Remarks: Numerous specimens occur without the distinctive bulla-like elongation of the final chamber. These specimens may represent a separate species or subspecies, but are more likely immature or ecological variants.

Distribution: PARKER (1962) reports it between 20°–65°S in the Pacific and BRADSHAW (1959) mapped it north of 30°N in the Pacific.

Plate 10   *Globigerina quinqueloba* NATLAND, 1938   49

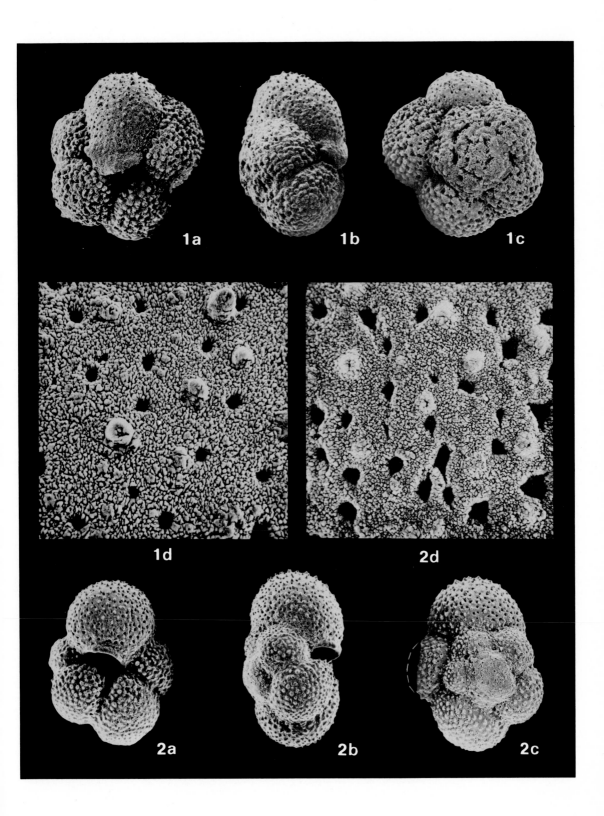

11. 1 *Globigerina rubescens* HOFKER, 1956
Plate 11, Figs. 1a–c, × 266; 1d, × 1066

Globigerina rubescens HOFKER, 1956, p. 234, pl. 32, figs. 18–21. — PARKER, 1962. pl. 2, Fig. 17–18.
Globigerina rosacea BERMÚDEZ and SEIGLIE, 1963, pp. 69–70, pl. 29, figs. 1–3.
"Globigerina" rubescens rubescens (HOFKER). — FLEISHER, 1974a, p. 1020, pl. 6, fig. 9; pl. 7, fig. 2.

Types: Syntype (*rubescens*)—Malaya-Indonesian Archipelago area, found in deep water.

Syntype (*rosacea*)—the Cariaco Basin, north of Venezuela, 703 m.

Plesiotype—V24-150, trigger core top (Lat. 02° 12′S, Long. 155° 42′E, 1,849 m).

Diagnosis: Test small, low to medium height trochospire, 4 globular chambers in final whorl, about 11–15 chambers arranged in about 3–3$^1/_2$ whorls. In recent and late Pleistocene sediments it has distinctive pink to red pigmentation. Chambers subspherical, closely packed although sutures are distinct, increasing rapidly in size as added. Aperture umbilical, a high open arch with an imperforate rim-like lip. Wall calcareous, coarsely perforate, with wide polygonal interpore cover and small pores with funnel-like depressions, although rarely inspectable on the apertual face, spinose. Spines are simple circular cross-section spines.

Remarks: The most distinct feature of this species is its pink to red pigmentation. The chemical nature of this pigment has not been determined. In Early Pleistocene sediments this species has been observed to be white possibly owing to a diagenetic loss or biochemical decomposition of this pigment. This form can be distinguished from *G. ruber* pink variety and *G. tenellus* by its more lobulate periphery and lack of supplementary apertures.

Distribution: Early Pleistocene (N.21) to Recent, temperate to equatorial waters. See also FRERICHS, 1968, pp. 1456–1458 for Pleistocene climatic variations.

11. 2 *Globigerinoides tenellus* PARKER, 1958
Plate 11, Figs. 2a–c, × 266; 2d, × 1066

Globigerinoides tenella PARKER, 1958, p. 280, pl. 6, figs. 7–11.
Globigerinoides tenellus PARKER. — PARKER, 1962, p. 232, pl. 4, figs. 11–12.
"Globigerina" tenella (PARKER). — FLEISHER, 1974a, p. 1020, pl. 6, figs. 5, 6; pl. 7, fig. 3.

Types: Holotype—Atlantis Station 4711 (Lat. 34° 58′N, Long. 19° 24′E, 3,309 m).

Plesiotype—V24-53, trigger core top (Lat. 01° 51′N, Long. 129° 01′N, 4,473 m).

Diagnosis: Test small, low to medium height trochospire, 4 globular chambers forming the final whorl, about 10 chambers arranged in 3 whorls. Chambers subspherical to ovoid, much embracing without distinct sutures, rapidly enlarging. Primary aperture umbilical, an open and high arch with a faint rim-like lip; secondary aperture(s) visible on spiral side intersection of spiral and intercameral sutures. Wall calcareous, coarsely perforated, pores set in weakly developed polygonal pore depressions, spinose. Spines are simple rounded spines with small upraised spine bases.

Remarks: This form can be distinguished from *G. rubescens* by its dorsal supplementary aperture(s), visible on even the smallest specimens, and from *G. ruber* by its more lobulate periphery.

Distribution: Late Pleistocene (N.21) to Recent, equatorial to temperate waters.

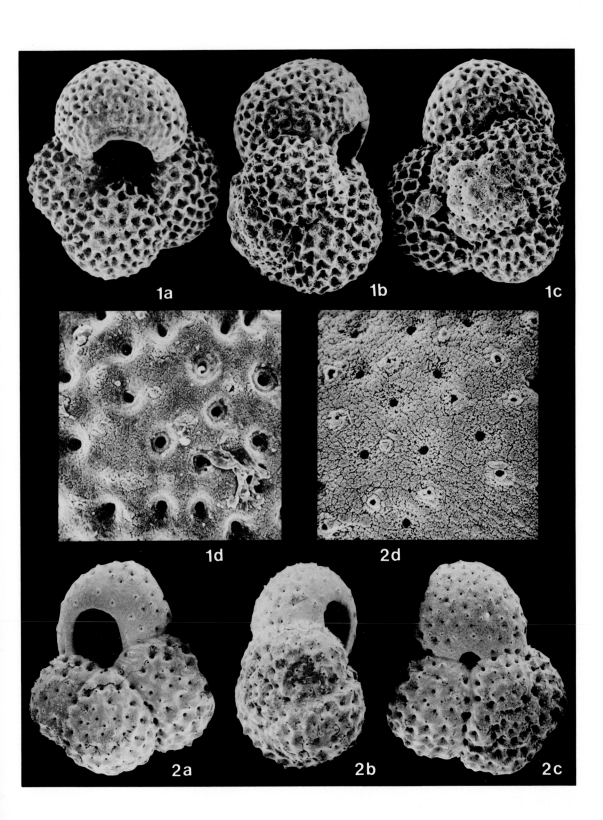

1a   1b   1c

1d   2d

2a   2b   2c

12. 1 *Globigerina decoraperta* TAKAYANAGI and SAITO, 1962
Plate 12, Figs. 1a–c, × 240; 1d, × 1600

Globigerina druryi AKERS subsp. *decoraperta* TAKAYANAGI and SAITO, 1962, p. 85, pl. 28, figs. 10a–c.
Globigerina decoraperta TAKAYANAGI and SAITO — PARKER, 1967, p. 149, pl. 19, figs. 1a–c, not fig. 2.
 — BRÖNNIMANN and RESIG, 1971, pl. 1293, pl. 6, fig. 2.
"Globigerina" rubescens decoraperta (TAKAYANAGI and SAITO). — FLEISHER, 1974a, pp. 1019–1020, pl. 6,
 fig. 8; pl. 7, fig. 4.

Types: Holotype—Sample A-16, Nobori Fm., Nobori, Muroto City, Kochi Prefecture,
 Japan (Lat. 33°22′09″N, Long. 134°03′33″E).
 Topotype—Sample A-6, Nobori Fm.

Diagnosis: Test small, low to medium height trochospirel, 4 globular chambers in
the final whorl, about 12 chambers in all arranged in about 3 whorls. Chambers spherical, closely packed, increasing rapidly in size, last chamber overhanging umbilicus slightly, sutures distinct. Aperture interiomarginal-umbilical, a medium-height, symmetrical
arch with a prominent rim-like lip. Wall calcareous, finely perforated, with widely-
spaced pores in funnel-shaped depressions, spinose. Spines (as inferred from spine
bases) simple round spines growing from low spine bases.

Remarks: This species can be distinguished from *G. tenellus* by its lack of supplementary apertures, and from *G. rubescens* by its lack of pink pigmentation. *G. decoraperta*
has, however, been observed in Middle Pleistocene sediments from the equatorial Pacific accompanying white (bleached?) *G. rubescens*. *G. decoraperta* was here separated
by its thicker lip and slightly larger aperture, and SEM photos suggest that it also has
finer perforations and a smoother surface.

Distribution: Middle Miocene (N. 14) to Late Pleistocene (N. 22), probably in tropical to temperate waters.

12. 2 *Berggrenia riedeli* RÖGL and BOLLI, 1973
Plate 12, Figs. 2a–b, 3, × 480; 2c, × 3200

Hastigerinella riedeli RÖGL and BOLLI, 1973, p. 507, pl. 4, figs. 1–5, pl. 14, figs. 1–3, text-figs. 5a–b.
Hastigerinopsis riedeli (RÖGL and BOLLI). — POORE, 1979 p. 472, pl. 19, figs. 1–4.

Types: Holotype—DSDP Leg 15, Site 147, Core 6, core catcher (Lat. 10°42′48″N,
 Long. 65°10′48″W, 51 m).
 Near-Topotype—V12-98, 0–10 cm (Lat. 10°47. 1′N, Long. 65° 06.8′W, 736 m).

Diagnosis: Test very small, low height trochospire, 4 $1/2$–5 subglobular chambers in
the whorl, about 10–11 chambers in all arranged in about 2 $1/2$ whorls. Chambers subspherical to slightly radially elongate, increasing slowly in size as added, partly embracing, creating a wide but relatively flat test, sutures depressed; last chamber tilted toward
the umbilicus and often slightly covering it. Aperture interiomarginal-umbilical, the last
chamber usually has a thin lip. Wall calcareous, thin, very finely and irregularly perforated, mostly near the peripheral ends of the chambers, triangular possibly becoming
triradiate spines in stout rounded spine bases.

Remarks: Based on the previous work of SAITO, THOMPSON and BREGER (1976), the
chamber shape, spine base and spine morphology of this species are not typical of *Hastigerinella*. We have provisionally assigned it to *Berggrenia*.

Distribution: Late Pleistocene to Recent, reported only in the Cariaco Basin.

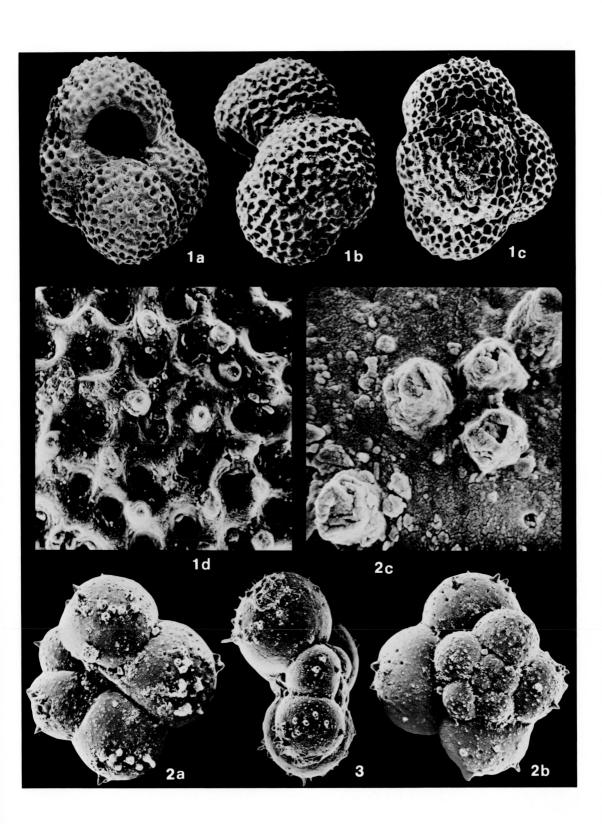

13. 1   *Globigerina antarctica* KEANY and KENNETT, 1972
Plate 13, Figs. 1a, 2a–b, × 240; 1b, × 2400

Globigerina antarctica KEANY and KENNETT, 1972, pp. 532–533, text-fig. 3, nos. 1–6.

Types: Holotype—Eltanin 36–14, 500–502 cm (Lat. 58°06′S, Long. 150°10′E, 3,054 m).
Plesiotype—Eltanin 36–14, 60 cm.

Diagnosis: Test medium in size, 4 globular chambers in the final whorl, about 12–14 chambers in all arranged in a low trochospire of about 2 ¹/₂ whorls, chambers sub-spherical to spherical, size increasing slowly as added, chambers well separated giving a fairly lobulate equatorial peripheral outline, sutures depressed, radial. Aperture umbilical, a moderately high arch bordered by a thin rim-like lip. Wall calcareous, very finely perforated, probably spinose although the last chamber tends to be smooth.

Remarks: KEANY and KENNETT distinguished this form from *G. bulloides* by its greater lobulation, distinct apertural lip, and smoother test wall, and from *G. falconensis* by being less lobulate, having a smoother wall and a high arched aperture. It may also be closely related to *G. megastoma*.

Distribution: Reported from the Matuyama Magnetic Epoch of the early and middle Pleistocene, and is most abundant south of the Antarctic Convergence.

13. 2   *Globigerinoides obliquus* BOLLI, 1957
Plate 13, Figs. 3a–c, × 160; 3d, × 1600

Globigerinoides obliqua BOLLI, 1957, pp. 112–114, pl. 25, figs. 9a–10c; text-fig. 21, no. 5.
Globigerinoides obliquus BOLLI. — BLOW, 1969, p. 324. — AKERS, 1972, p. 60, pl. 7, figs. 1a–c; pl. 10, figs. 1a–d; pl. 16, figs. 1a–b; pl. 18, figs. 2a–b; pl. 26, figs. 2a–b; pl. 27, figs. 3a–c; pl. 35, figs. 1a–b; pl. 53, figs. 2a–b; pl. 56, figs. 1a–b; pl. 58, figs. 3a–b. — JENKINS and ORR, 1972, p. 1092, pl. 14, figs. 8–10. — BOLTOVSKOY and WATANABE, 1975, pp. 40–41, text-figs. 1a–2b.

Types: Holotype—BOLLI's sample KR23422 (TTOC 160634), type locality of the
Globorotalia mayeri Zone (Miocene), Trinidad.
Plesiotype—2.5 m above the Pliocene–Pleistocene boundary in the Le Castella type section, near Crotone, southern Italy.

Diagnosis: Test medium to large in size, 3–4 globular chambers in the final whorl, about 12–15 chambers in all arranged in a medium height trochospire of about 3 whorls, chambers sub-spherical becoming gradually more radially flattened as added, size increases rapidly through ontogeny although the final chamber is often reduced in size, sutures radial, depressed. Primary aperture umbilical, interiomarginal, a high and wide arch with a thin rim-like lip; one or more supplementary apertures on the dorsal side also quite large. Wall calcareous, perforated, rounded spines.

Remarks: SAITO, BURCKLE and HAYS (1975, pp. 234–235) reported *G. obliquus* as occurring slightly above the Pliocene-Pleistocene boundary at the Le Castella section as well as in deep-sea piston cores. It has been noted (THOMPSON and SCIARRILLO, 1978) that this species is dissolution susceptible and it is often difficult to locate the exact extinction level.

Distribution: Middle Miocene to earliest Pleistocene, but also reported from the Recent of the South Pacific by BOLTOVSKOY and WATANABE (1975). Ecology is probably similar to the modern form *G. conglobatus*.

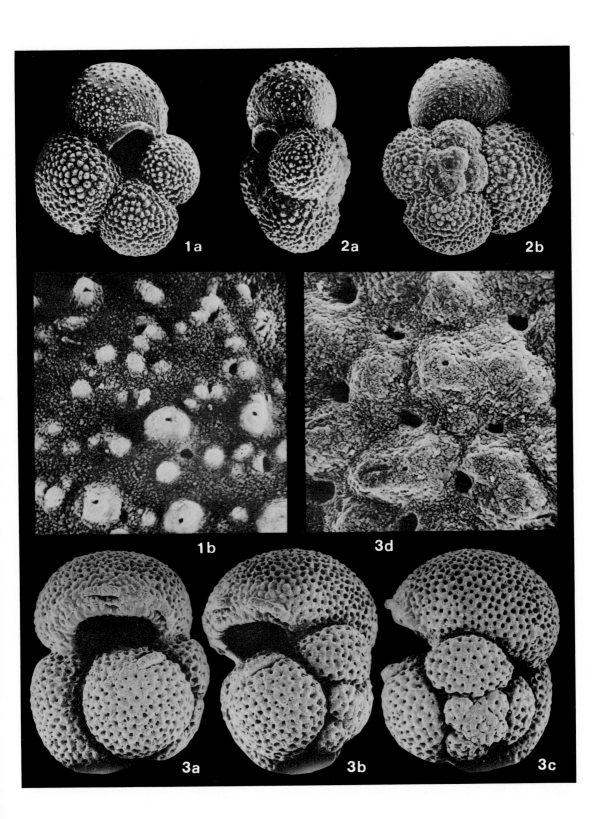

14. 1 *Globigerinoides conglobatus* (BRADY), 1879
Plate 14, Figs. 1a–c, × 71; 1d × 535

Globigerina conglobata BRADY, 1879, p. 286, (no figures). — BRADY, 1884, p. 603, pl. 80, figs. 1–5; pl. 82, fig. 5. — BANNER and BLOW, 1960a, pp. 6,7, pl. 4, fig. 4 (lectotype).
Globigerina dutertrei D'ORBIGNY. — CUSHMAN, 1922, (not D'ORBIGNY, 1839), p. 36, pl. 5, figs. 8, 9.
Globigerina (Globigerinoides) conglobata BRADY. — COLOM, 1952, p. 42.
Globigerinoides conglobata (BRADY). — PARKER, 1948, p. 238, pl. 7, figs. 8,9.
Globigernoides conglobatus (BRADY). — DROOGER, 1953, p. 142 — A.G.I.P. Mineraria Auct., 1957, Tav. 46, fig. 1.
Globigerinoides canimarensis BERMÚDEZ, 1961, p. 1225, pl. 10, figs. 5a–b.
Globigerinoides conglobatus canimarensis BERMÚDEZ. — BLOW, 1969, p. 324, pl. 20, figs. 8; pl. 21, fig, 1 (Lat. 35°35'N, Long 50°27'W, 2,750 fm).

Types: Syntypes—Challenger Station 64 and Challenger Station 338 (Lat. 21° 15'S, Long. 14°02'W, 1,990 fm).

Lectotype—Challenger Station 338, BRADY, 1884, pl. 80, fig. 1, by BANNER and BLOW, 1960a, p. 6.

Neartopotype—RC8-23, trigger core top (Lat.25°09'S, Long.12°46'W, 3,338m).

Diagnosis: Test large, medium to high trochospire, about 3 1/2 subglobular chambers in the final whorl, about 10 chambers arranged in about 3 whorls. Chambers initially subspherical becoming radially flattened, rapidly enlarging as added but greatly embracing so that the test becomes subglobular or subquadrate in outline. Sutures very deep and distinct. Primary aperture interiomarginal, umbilical, a long, low asymmetrical arch with a thin perforated rim-like lip; secondary supplementary apertures on spiral side at intersections of spiral and intercameral sutures, also with apertural rims. Wall calcareous, thick, coarsely perforated with pores set in funnel-like depressions having wide and flat interpore areas, spinose. Spines are round to rounded-trinagular in simple spine bases set in large upraised spine bases.

Remarks: This species can be distinguished from *G. ruber* by its larger size in adults, and more embracing chambers in juveniles. It is also less compact and slightly more lobular than *G. gomitulus*.

Distribution: Latest Miocene (N. 18) to Recent, in equatorial to temperate waters.

14. 2 *Globigerinoides gomitulus* (SEGUENZA), 1880
Plate 14, Figs. 2a–c, × 71, 2d, × 535

?"*Polymorphium globuliferum*" (teste D'ORBIGNY, 1826) SOLDANI, 1791, p. 119 (? not p. 118), pl. 130, figs. qq, rr?, pp?
Globigerina gomitulus SEGUENZA 1880, p. 308, pl. 17, figs. 16, 16a.
Globigerinoides gomitulus (SEGUENZA). — MISTRETTA, 1962, pp. 100–102, p. 101, text-figs. 1a–c (neotype).
Globigerina helicina D'ORBIGNY, 1826, p. 277, list no. 5 (no figs.). — FORNASINI, 1899, p. 209, text-fig. 4 (after D'ORBIGNY). — PARKER, JONES and BRADY, 1871, pl. 11, fig. 113 (after SOLDANI, 1791, fig. qq). — FORNASINI, 1899, p. 209, fig. 4. — BANNER and BLOW, 1960a, pp. 13–14, pl. 3, fig. 5 (lectotype).
Globigerina adriatica FORNASINI, 1898 (part), p. 582, pl. 3, ?fig. 7.

Types: Syntype—Riace, Province of Vessia, Calabria, Italy (Pliocene).
Neotype—Same locality.
Plesiotype—RC 10-114, 1,560 cm (11°11'S, 162°55'W, 2,791 m).

Diagnosis: Test size medium to large, a medium height trochospire, 3–4 subglobular

Plate 14 *Globigerinoides conglobatus* (Brady), 1879
Globigerinoides gomitulus (Seguenza), 1880

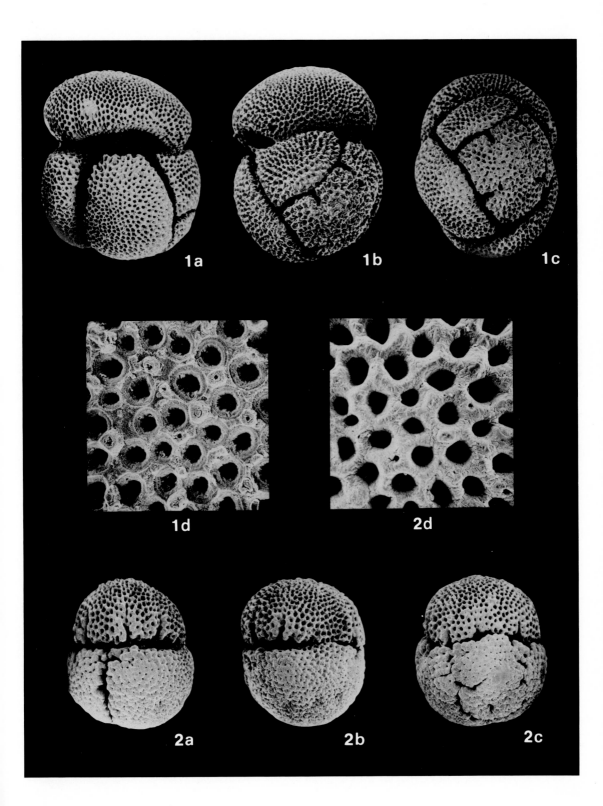

chambers in the final whorl, about 10 chambers in all. Chambers initially subspherical, becoming radially flattened, which, combined with the greatly embracing nature of the chambers gives the test a very round periphery; sutures distinct, deeply incised. Primary aperture probably umbilical, a very low sutural opening with no apparent lip; secondary apertures also sutural at the intersections of spiral and intercameral sutures. Wall calcareous, very thick, densely perforated with large irregularly shaped pores, probably spinose. Spines probably simple rounded spines.

Remarks: This form is similar to *G. conglobatus* and is probably phylogenetically related to it, but can be differentiated by the closer packing of the chambers and smaller size. Specimens from the plesiotype locality possess a light pink coloration. MISTRETTA (1962) considered it a junior synonym of *G. conglobatus*.

Distribution: Pleistocene, probably occupying similar conditions as *G. conglobatus*.

15. 1 *Globigerinoides ruber* (D'ORBIGNY), 1839
Plate 15, Figs. 1a–c, × 100; 1d, × 1000
(see also Pl. 51, Fig. 3)

Globigerina rubra D'ORBIGNY, 1839a, pp. 82–83, (plates published separately), pl. 4, figs. 12–14. — BANNER and BLOW, 1960a, p. 19, pl. 3, figs. 8a–b (lectotype).
Pylodexia rubra EHRENBERG, 1873, p. 293.
Globigerina bulloides D'ORBIGNY var. *rubra* D'ORBIGNY. — VAN DEN BROECK, 1876, p. 125, pl. 3, figs. 7, 9,10.
Globigerinoides rubra (D'ORBIGNY). — CUSHMAN, 1927, p. 87. (not pl. 19, fig. 6). — BOLLI, 1957 (part), pp. 113–114, pl. 25, figs. 12a–c (not fig. 13).
Globigerinoides ruber (D'ORBIGNY). — PARKER, 1962, p. 230, pl. 3, figs. 11–14; pl. 4, figs. 1–10. — ORR, 1969, pp. 373–379. pl. 1.
Globigerinoides ruber (D'ORBIGNY) forma *alba* — BOLTOVSKOY, 1968, pp. 89–90.
Globigerinoides ruber (D'ORBIGNY) forma *rosea*. — BOLTOVSKOY, 1968, pp. 89–90, pl. 1, figs. 2,3.

Types: Syntypes—listed as occurring in the (marine) sands of Cuba, Jamaica, Guadaloupe and Martinique.

Lectotype—D'ORBIGNY's material from Cuba.

Plesiotype—V18-10, trigger core top (Lat. 17°07'N, Long. 65°02'W, 4,430 m).

Diagnosis: Test small to large in size, low trochospiral coil, 3 globular chambers in the final whorl, about 10 chambers arranged in about 2 whorls. May have pink-to-red pigmentation. Chambers spherical to slightly flattened radially, increasing slowly in size, slightly separated. Primary aperture interiomarginal, umbilical. A small to moderately high arch sometimes with a slight lip, secondary aperture(s) on spiral side at intersection(s) of spiral and intercameral sutures. Wall calcareous, coarsely perforated, spinose. Spines with rounded cross-section, spines set in slightly raised bases.

Remarks: D'ORBIGNY originally described this species on the basis of the red-to-pink variety he noted in the Caribbean, stating that it also could be found as yellowish red or yellow. Two varieties are noted in deep-sea sediments, pink and white, unless some staining has occurred. Further, the white variety is far more abundant in both quantity and occurrence than the pink form. The pink form, however, has not been found in surface sediments of the Indo-Pacific area, although it is found in Late Pleistocene core sediments. Many specimens observed in the Indo-Pacific area which could be considered pink varieties have an early whorl or two of light pink chambers and gradually become white in later chambers. BÉ and HAMLIN (1967) have suggested that the pigmentation is due to the volatile chemical pheophytin and ORR (1969) has added that this may be related to the zooxanthellae activity within the foraminifera's protoplasm. Near the edge of its distributional area in the Pacific, the white variety becomes very small and the chambers are very closely packed. *G. ruber* can be distinguished from *G. cyclostoma* by its (typically) larger size and slightly more lobulate periphery. *G. helicina* (see Pl. 56, Fig. 7) is here considered a junior synonym of *G. ruber* based on SEM observation of this aberrant form.

Distribution: Late Miocene (N. 16) to Recent, equatorial to temperate waters. CORDEY (1967) documented in detail the phylogenetic development of *G. ruber* from the Miocene to Recent interval. Thompson *et al.* (1979) placed the extinction level of this species with pink-pigmented tests in the Indo-Pacific at 120,000 years B.P.

15. 2 *Globigerinoides cyclostomus* (GALLOWAY and WISSLER), 1927
Plate 15, Figs. 2a–c, × 100; 2d, × 1000

Globigerina cyclostoma GALLOWAY and WISSLER, 1927, p. 42, pl. 7, figs. 8–9.
Globigerinoides cyclostomus (GALLOWAY and WISSLER). — ASANO, 1957, pp. 23–24, pl. 1, figs. 22,23.

Types: Holotype and topotype—Lower bed at the Lomita Quarry, Palos Verdes Hills, near Los Angeles, California; collected by A. R. LOEBLICH, JR.

Diagnosis: Test small, low to medium height trochospire, 3 globular chambers in the final whorl, about 11 chambers arranged in about 3 whorls. Chambers subspherical to ovoid, increasing moderately in size but closely packed giving the test a rectangular outline. Primary aperture umbilical, a small oval opening with no modification, secondary aperture(s) on spiral side at intersection of spiral and intercameral sutures. Wall calcareous, coarsely perforated, spinose. Spines simple rounded cross-section, spines in simple raised bases (spines not observed on specimens examined, but bases preserved).

Remarks: This form has been reported from the Pleistocene sediments of California and Japan, and is rarely found in Pacific deep-sea sediments of temperate latitudes and may simply be a local variation of *G. ruber*. It can be distinguished from *G. ruber* by its more compact chamber arrangement and small aperture relative to chamber size.

Distribution: Pleistocene (?) to Recent in temperate waters.

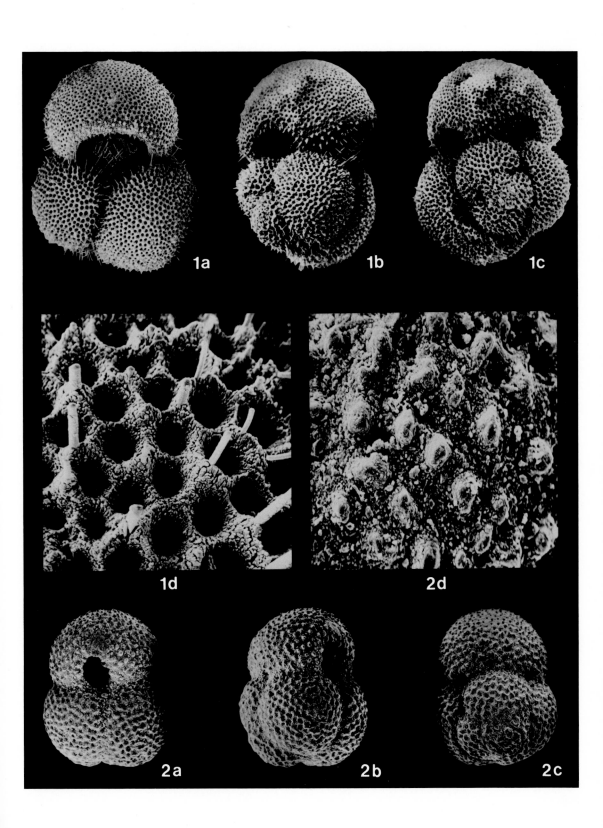

16. 1   *Globigerinoides pyramidalis* (VAN DEN BROECK), 1876
Plate 16, Figs. 1a–c, × 107; 1d, × 1250

Globigerina bulloides D'ORBIGNY var. *rubra* D'ORBIGNY subvar. *pyramidalis* VAN DEN BROECK, 1876, p. 127, pl. 3, figs. 9–10.

Globigerina rubra D'ORBIGNY. — BRADY, 1884, pl. 79, figs. 13–15.

Globigerina pyramidalis (VAN DEN BROECK). — BANNER and BLOW, 1960a, p. 21.

Globigerinoides ruber var. *pyramidalis* (VAN DEN BROECK). — BERMÚDEZ, 1961, pp. 1235–1236, pl. 11, figs. 2, 3a–c.

Globigerinoides ruber (D'ORBIGNY) forma *pyramidalis* — BOLTOVSKOY, 1968, pp. 89–90.

Globigerinoides elongata cedrosensis McCULLOCH, 1979, p. 418, pl. 174, figs. 1, 3 (?).

Types: Holotype——Gulf of Gascony, France.

Plesiotype——RC8-27, trigger core top (Lat. 26°14′S, Long. 10°36.5′W, 3,678 m).

Diagnosis: Test medium to large, very high trochospire, 3 subglobular chambers in the final whorl, about 12–15 chambers arranged in about 3 whorls. Chambers initially subspherical becoming flattened radially, size increases rapidly but coiling is tight giving rise to its typical pyramid shape, sutures deep. Primary aperture large, wide, rounded arch, nearly as big as the final chamber which is usually dwarfed. Secondary supplementary apertures on spiral side also large, circular, formed at intersection of spiral and intercameral sutures. Wall calcareous, uniformly and coarsely perforate, spinose. Spines (not found intact—inference from spine base) round to rounded-triangular cross-section in simple raised spine bases.

Remarks: This species is easily distinguished from *G. ruber* or *G. elongatus* by its high trochospire and prominent secondary apertures.

Distribution: Reported only from latest Pleistocene and Recent sediments and equatorial waters.

16. 2   *Globigerinoides elongatus* (D'ORBIGNY), 1926
Plate 16, Fig. 2a–c, × 107; 2d, × 1250

(?)"*Polymorphium tuberosum et globiferum*" (teste D'ORBIGNY, recte. "*polym. tuberosa et Globuliferae*") SOLDANI, 1791, p. 117, tab. 123, fig. K (not figs. H,I,J,L,M,N,O,P).

Globigerina elongata D'ORBIGNY, 1826, p. 277, list no. 4, (no figs.) — PARKER, JONES and BRADY, 1871, pl. 11, fig. 129 (after SOLDANI, fig. K). — FORNASINI, 1899, p. 207, fig. 1. — BANNER and BLOW, 1960a, pp. 12–13, pl. 3, fig. 10a–c (lectotype).

Globigerina canariensis D'ORBIGNY, 1839b, p. 133, pl. 2, figs. 10–12. — LE CALVEZ, 1974, pp. 17–18.

Globigerinoides elongata (D'ORBIGNY). — CUSHMAN, 1941, p. 40, pl. 10, fig. 20–23; pl. 11, fig. 3.

?Globigerinoides elongatus (D'ORBIGNY). — A.G.I.P. Mineraria Auct, 1957, pl. 46, figs. 2d, 2v.

Globigerinoides ruber (D'ORBIGNY) forma *elongata* — BOLTOVSKOY, 1968, pp. 89–90, pl. 1, fig. 4.

Types: Syntypes—Near Rimini, on the Adriatic Sea.

Lectotype—Specimen selected from D'ORBIGNY's remaining syntypes by BANNER and BLOW, 1960a, p. 12.

Plesiotype—Caryn 22-6, core top (Lat. 32°16′N, Long. 64°38.75′W, 1,509 m).

Diagnosis: Test medium to large, low to medium height trochospire with 3 subglobular chambers in the final whorl, about 10 chambers arranged in about 3 whorls. Chambers ovoid to rectangular, tightly packed, increasing slowly in size to give the test

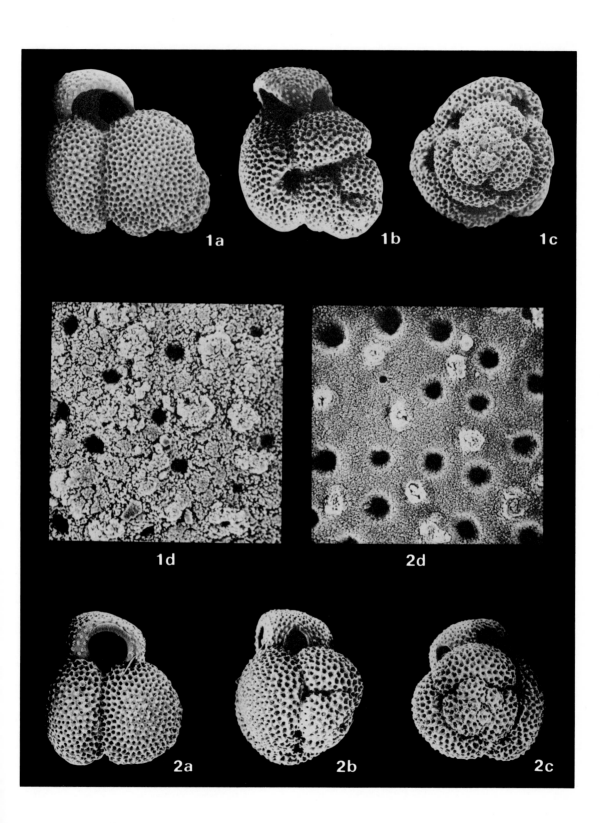

a "sub-tetrahedroid" shape. Sutures distinctly depressed. Primary aperture a large circular opening almost as high and wide as the final chamber, often with a noticeable imperforate rim-like lip. Secondary apertures on spiral side highly arched rounded openings at intersection of spiral suture and intercameral sutures. Wall calcareous, thick, coarsely perforated with polygonal, funnel-like pore depressions, spinose. Spines round to rounded-triangular in low, raised spine bases.

Remarks: *G. elongatus* can be distinguished from *G. ruber* and *G. pyramidalis* by its more tightly coiled trochospire of less-inflated chambers and thicker wall.

Distribution: Late Miocene (N. 16) to Recent, probably in similar (although more rarely) conditions as *G. ruber*.

17   *Globigerinoides sacculifer* (BRADY), 1877
Plate 17, Figs. 1a–c, 2a–c, × 90; 1d, 2d, × 600
(see also Pl. 56, Figs. 7a–c)

Globigerina helicina D'ORBIGNY. — CARPENTER, 1862 (not D'ORBIGNY), pl. 12, fig. 11.
Globigerina sacculifera BRADY, 1877, p. 535 (no figs.). — BRADY, 1884, p. 604, pl. 80, figs. 11–17; pl. 81, fig. 2; pl. 82, fig. 4. — BANNER and BLOW, 1960a, p. 21, pl. 4, figs. 1 (lectotype), 2.
Globigerina tricamerata TOLMACHOFF, 1934, p. 324, pl. 41, fig. 21.
Globigerina sacculiferus (BRADY). — LEROY, 1941, p. 44, pl. 2, figs. 68–70; p. 87, pl. 4, figs. 25–27.
Globigerinoides triloba sacculifera (BRADY). — BOLLI, 1957, p. 113, pl. 25, fig. 5a–6.
Globigerina bulloides D'ORBIGNY var. *recumbens* RHUMBLER, 1901, p. 25, text-fig. 27.
Globigerina sacculifera BRADY var. *recumbens* RHUMBLER, 1911 (1949), pl. 31, figs. 11–13.
Globigerina sacculifera BRADY var. *galeata* RHUMBLER, 1911 (1949), pl. 31, figs. 14–15.
Globigernoides suleki BERMÚDEZ, 1961, pp. 1241–1242, pl. 10, fig. 9a–b. — RÖGL and BOLLI, 1973, p. 565, pl. 4, fig. 23; pl. 15, fig. 3.
Globigerinoides quadrilobatus sacculifer (BRADY). — PARKER, 1962, p. 229, pl. 3, figs. 6–10.
Globigerinoides sacculifer (BRADY). — PARKER, 1967, pp. 156–158, pl. 21, figs. 1,2,4, text-fig. 5.
Globigerinoides trilobus (REUSS) forma *typica*. — BOLTOVSKOY, 1968, p. 59, pl. 1, fig. 1.
Globigerinoides trilobus (REUSS) forma *sacculifera*. — BOLTOVSKOY, 1969, p. 89, pl. 2, fig. 23.
Globigerinoides immatura LEROY. — BERMÚDEZ, 1960, p. 72, pl. 1, figs. 35,36.
Globigerinoides triloba immatura LEROY. — BERMÚDEZ and BOLLI, 1969, pp. 162–163, pl. 7, fig. 7.
Globigerinoides sacculifera (BRADY). — CUSHMAN and JARVIS, 1930, p. 366, pl. 34, fig. 4.
Globigerina bulloides D'ORBIGNY var. *triloba* REUSS. — BRADY, 1884, p. 595, pl. 79, fig. 2.
Globigerinoides triloba (REUSS) — DROOGER and KAASSCHIETER, 1958, p. 83, pl. 5, fig. 8.
Globigerinoides triloba triloba (REUSS) — BERMÚDEZ and BOLLI, 1969, pp. 165–166, pl. 9, figs. 7–9.

Types: Holotype (*sacculifera*)—From "chalk" on a beach near Liversidge in New Ireland territory, New Guinea. (Probably Upper Miocene or Pliocene according to BANNER and BLOW, 1960a, p. 21.)

Lectotype—Challenger Station 224 (Lat. 07°45′N, Long. 144°20′E, 1,850 m).

Holotype (*suleki*)—Atlantis Station 2988 (Lat. 23°15′N, Long. 79°57′W, 380 fm).

Plesiotype (*sacculifer* without "sack")—V3-3, 0–1 cm (Lat. 18°51′N, Long. 67°07′W, 2,661 m); (*sacculifer* with "sack")—V18-10, 0–1 cm (Lat. 17°07′N, Long. 65°02′W, 4,430 m).

Diagnosis: Test medium to large in size, low trochospire, 3–4 globular chambers in the final whorl, about 10 chambers arranged in about 2 $1/2$ whorls. Chambers spherical to slightly flattened radially, very rapidly enlarging, partially embracing with distinct sutures. Final chamber may be incompletely inflated and have the appearance of a "wine-sac." Primary aperture interiomarginal, umbilical, a low but quite wide symmetrical arch; supplementary apertures on spiral side usually elliptical or sub-triangular, formed at intersections of intercameral and spiral sutures. Wall calcareous, coarsely perforated with large, deep, steep-walled pores with narrow interpore areas, spinose. Spines round to rounded triangular set in fairly high spine bases.

Remarks: Both forms with and without the typical kummerform sac occur commonly in Pleistocene and Recent warm waters. The pair may represent separate species or subspecies but may also be ecologic variants. Comparison of the two examples figured show no basic differences other than a secondary cortex on the lower figure. TOLMACHOFF

(1934) placed BARDY's (1884, pl. 81, figs. 2–3) Recent variety *"triloba"* in synonymy with his new Miocene species. Lobulate forms without sacs are often referred to the Miocene species *Globigerina trilocularis* D'ORBIGNY, *Globigerina triloba* REUSS or *Globigerinoides immatura* LEROY.

Plate 17   *Globigerinoides sacculifer* (Brady), 1877   67

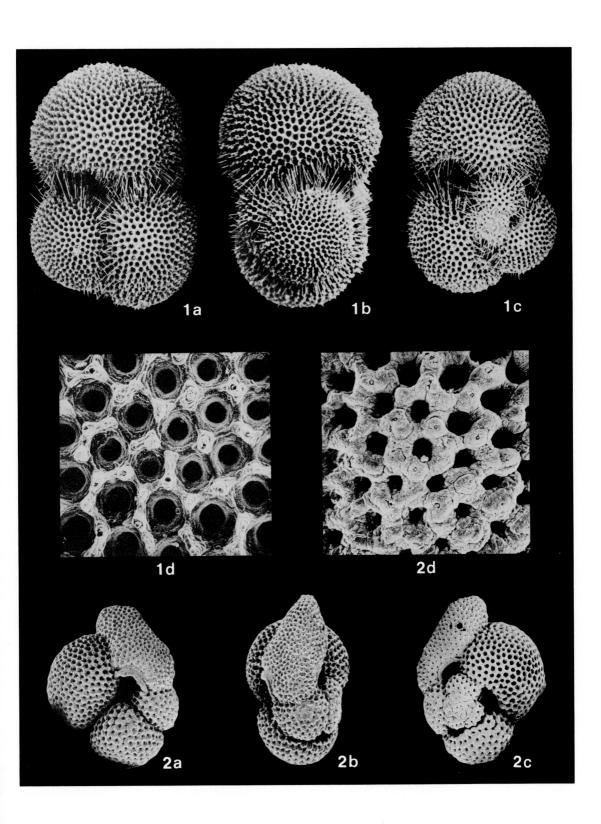

18. 1 *Globigerinoides fistulosus* (SCHUBERT), 1910
Plate 18, Figs. 1a, 2–3, × 50; 1b, × 666

Globigerina fistulosa SCHUBERT, 1910, pp. 323, 324, text-fig. 2. — SCHUBERT, 1911, p. 10, text-fig. 13a–c.
Globigerinoides sacculifera (BRADY) var. *fistulosa* (SCHUBERT). — CUSHMAN, 1933, pl. 34, figs. 6a–c.
Globigerinoides quadrilobatus (D'ORBIGNY) *fistulosus* (SCHUBERT). — BELFORD, 1962, pp. 16–17, pl. 4, figs. 7–10.
Globigerinoides quadrilobatus (D'ORBIGNY) *hystricosus* BELFORD, 1962, p. 17, pl. 4, figs. 11–14.
Globigerinoides sacculifer fistulosa (SCHUBERT). — TODD, 1964, p. 1084, pl. 290, fig. 6.
Globigerinoides trilobus fistulosus (SCHUBERT). — BOLLI, 1970, p. 579, pl. 1, figs. 8–11.
Globigerinoidesella fistulosa (SCHUBERT). — EL-NAGGAR, 1971, p. 476.
Globigerinoides fistulosus (SCHUBERT). — PARKER, 1967, pp. 154–155, pl. 21, figs. 3,5,6; text-fig. 4. — KENNETT, 1973, p. 5, figs. 14–15; pl. 6, figs. 2–11, p. 7, figs. 1–13; pl. 8, figs. 1–10.

Types: Syntype—Sandwich Island area, Bismarck Archipelago.
Plesiotype—RC13-78, 1,011 cm (Lat. 19°00.7'S, Long. 134°11.5'W, 2,725 m).

Diagnosis: Test size large to very large, a medium to high trochospire, 3–4 subglobular to pillow-shaped chambers in the final whorl, about 10–15 chambers arranged in 2 1/2 –3 whorls. Chambers initially spherical, becoming rapidly enlarged both radially and tangentially in the final whorl, with numerous distal protuberances on each chamber, sutures distinct, depressed. Primary aperture umbilical, a high and wide arch usually with a thin lip; secondary apertures located at intersections of intercameral and spinal sutures. Wall calcareous, densely perforated with circular pores, spinose. Spines probably similar to those of *G. sacculifer*.

Remarks: The bizarre morphology of this species makes identification easy. Early workers referred to the unique protuberances on the final whorl of chambers as "spines," but the spines of the *Globigerinoides* are now known to be very different in morphology. On some very large specimens, the wall of the test in the vicinity of the aperture grows to a point pointing inwards towards the umbilicus.

Distribution: Late Miocene (N. 18) to Early Pleistocene (N. 21). The extinction of this species is a good marker datum for the Olduvai Magnetic Event, occurring immediately after it. Its ecology is probably similar to *G. sacculifer*.

18. 2 *Globigerinoides sacculifer "hystricosus"* (BRADY), 1877—variant form
Plate 18, Figs. 4a–c, × 50; 4d, × 666
(see above for synonymy)

Types: Plesiotype—V26-127, 110 cm (Lat. 19°00'N, Long. 81°02'W, 6,251 m).

Remarks: This variant form of *G. sacculifer* is rare in occurrence but might be confused with *G. fistulosus*. It does not, however, have the extremely bizarre chamber shape and numerous protuberances of *G. fistulosus*, but simply represents an exotic form of *G. sacculifer*.

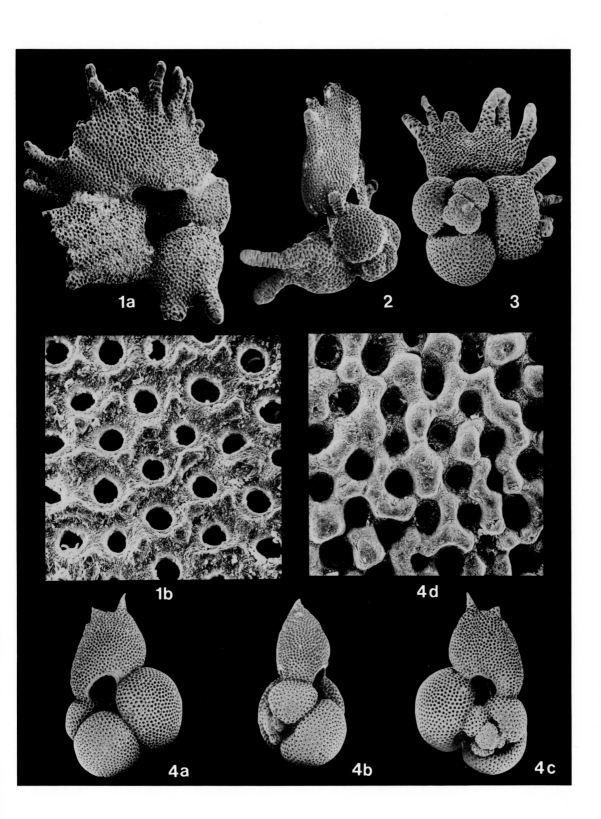

19 *Orbulina universa* (D'ORBIGNY), 1839

Plate 19, Figs. 1a–c, × 133; 1d, × 500; 2a, × 50; 2b, × 500; 3a, × 100; 3b, × 500; 4, 6, × 66; 5, × 50

Orbulina universa D'ORBIGNY, 1839a, p. 2, pl. 1, fig. 1. — CUSHMAN, 1914, pp. 14–15, pl. 6, figs. 1–5; pl. 7, figs. 1–2; pl. 11, fig. 3. — BELFORD, 1962, pp. 6–9, pl. 1, figs. 1–15.
Globigerina bilobata D'ORBIGNY, 1846, p. 164, pl. 9, figs. 11–14. — BANNER and BLOW, 1960a, pp. 2–3, pl. 3, fig. 9.
Globigerina (Orbulina) acerosa OWEN, 1867, p. 149, pl. 5, fig. 2.
Globigerina (Orbultna) continens OWEN, 1867, p. 149, pl. 5, figs. 3, 4, *Sphaeroides.*
Globigerina bulloides D'ORBIGNY var. *universa* D'ORBIGNY. — SIDDALL and BRADY, 1879, p. 7.
Globigerina (Sphaeroides) hastata EGGER, 1893, p. 368, pl. 13, fig. 80; pl. 14, figs. 41–42.
Orbulina universa D'ORBIGNY var. α, RHUMBLER, 1901, p. 28, text-fig. 29.
Orbulina universa D'ORBIGNY var. β, RHUMBLER, 1901, p. 28, text-fig. 30.
Orbulina imperfecta RHUMBLER, 1911, p. 218. — RHUMBLER, 1949, pl. 34, figs. 4,6.
Orbulina parva RHUMBLER, 1949, pl. 34, figs. 7, 8, 11, 12 (*nomen invalidum*).
Biorbulina bilobata BLOW, 1956, p. 69, text-fig. 2, fig. 16.

Types: (Trochoid form)—V28-301, trigger core top (Lat. 29°19′N, Long.128°54′E, 852 m).

(Spherical)—RC8-23, trigger core top (Lat. 25°09′S, Long. 12°46′E, 3,338 m).

(2–3 chambered forms)—Figs. 4–5 (V21-29, 25 cm: Lat. 00°57′ N, 89°21′ W, 712 m); Fig. 6 (V28-117, core bottom: Lat. 14°42′ N, 76°23′′ W, 4,027 m).

Diagnosis: Test large (spherical form), one (rarely 2 or 3) spherical chambers. Test small (trochoid form), moderate height trochospire, 4 chambers in the final whorl, about 13–15 globular chambers arranged in about 3 whorls, may remain visible on one side of spherical form. Chambers spherical in both forms. Trochoid form has chambers loosely embracing, rapidly enlarging, with deep sutures. Primary aperture (trochoid form only) interiomarginal, umbilical, partially blocked by overhanging ultimate chamber. Secondary supplementary apertures formed at intersection of spiral and intercameral sutures, frequently 2 occurring together with a thin wall between. Wall calcareous, thin, coarsely perforated with two distinct sizes of pores, spinose. Spines round-becoming-rounded-triangular or triradiate, simple round spine bases.

Remarks: The typically occurring form of this species is the large but fragile spherical form. The trochoid and two- or three-chambered spherical forms are rare and probably do not represent separate species. Bé, HARRISON and LOTT (1973) called attention to the different pore sizes, which, along with the overall test diameter, is latitudinally controlled in the Indian Ocean, and other oceans as well. The trochoid form may be distinguished from other species of *Globigerina* and *Globigerinoides* by its small size and large supplementary apertures. BLOW (1956) has proposed a separate bioseries for *Orbulina* and *Biorbulina,* although all variations of one-, two- and three- chambered varieties commonly occur in modern oceans.

Distribution: Early Miocene (N. 9) to Recent, in equatorial to temperate waters.

Plate 19   *Orbulina universa* (D'ORBIGNY), 1839   71

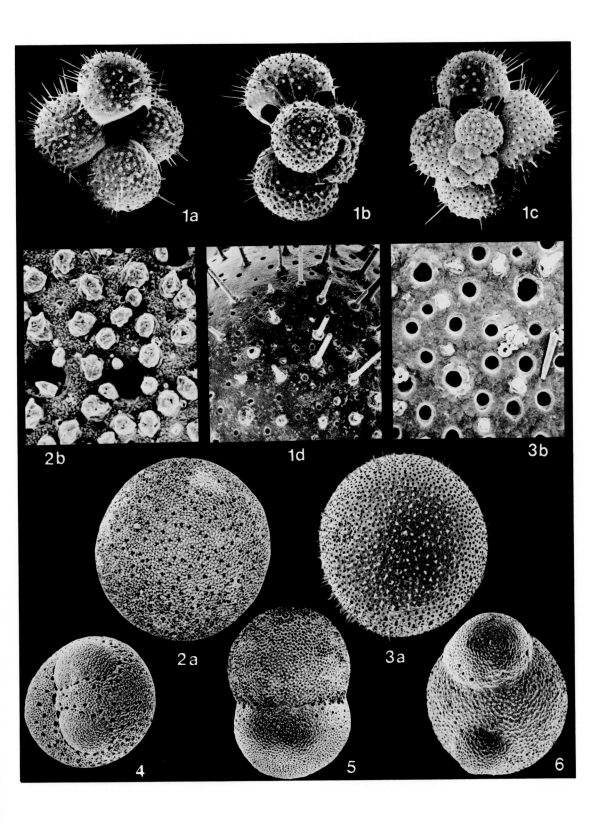

20. 1 *Sphaeroidinella dehiscens excavata* BANNER and BLOW, 1965
Plate 20, Figs. 1a–c, × 40; 1d, × 666

Sphaeroidina dehiscens (PARKER and JONES). — BRADY, 1884, pl. 84, fig. 8.
Sphaeroidinella dehiscens excavata BANNER and BLOW, 1965, p. 1164 (no figures). — BANNER and
    BLOW, 1967, p. 153, pl. 4, fig. 5 (holotype).

Types: Holotype—Challenger Station 224 (Lat. 07°45′N, Long. 144°20′E, 1,850 fm).
            Plesiotype—Mahi 70-PC44 (Lat. 13°14.7′S, Long. 160°27.2′E, 3,480 m).

Diagnosis: Test large, a medium-height trochospiral coil, about 4–5 irregularly shaped
chambers forming the last whorl, about 10–12 chambers in all. Chambers initially
spherical becoming greatly flattened radially with considerable embracement and deposition of secondary cortex rendering individual chamber shape quite indistinct. Chambers bordered with outward-turning chamber flanges; sutures very deep and wide, ex-
posing earlier chambers. Primary aperture umbilical, sutural, a wide opening following
the wide sutures partially separating the chambers of the last whorl. Secondary apertures
also sutural and usually connected through deep suturing. Wall calcareous, very thick,
densely perforated with large circular pores ringed with upraised collars, non-spinose
in adults. Spines may be present on juvenile forms, and probably are similar to those
possessed by *G. sacculifer.*

Remarks: BANNER and BLOW (1965) distinguished this form from *S. dehiscens, s.l.*
by the extreme suturing which exposes earlier whorls. They used it to define their N.22/
N.23 boundary, although this species is very rare in marine sediments.

Distribution: Latest Pleistocene (N. 23) to Recent, in habitats similar to *S. dehiscens.*

20. 2 *Sphaeroidinella dehiscens* (PARKER and JONES), 1865
Plate 20, Figs. 2a–c, × 66; 2d, × 666

Sphaeroidina bulloides D'ORBIGNY var. *dehiscens* PARKER and JONES, 1865, p. 369, pl. 19, figs. 5a–b. —
    BANNER and BLOW, 1960, p. 35, pl. 7, fig. 3.
Sphaeroidina dehiscens PARKER and JONES. — BRADY, 1884(part), p. 621, pl. 84, figs. 9–11. (not fig. 8).
Sphaeroidinella dehiscens (PARKER and JONES) — CUSHMAN, 1927, p. 90, pl. 19, fig. 2 (after PARKER
    and JONES). — BOLLI, LOEBLICH and TAPPAN, 1957, pp. 32–33, pl. 6, figs. 1–5.

Types: Syntypes—Material from "tropical Atlantic Ocean."
            Lectotype—From original syntypes (Lat. 02°20′N, Long. 28°44′W, 1,080 fm)
                by BOLLI, LOEBLICH and TAPPAN, 1957, p. 33; illustrated by BANNER and
                BLOW, 1960a, p. 35, pl. 7, fig. 3.
            Plesiotype—V9-17, trigger core top (Lat. 10°45.8′S, Long. 16°10.7′W, 4,186 m).

Diagnosis: Test large, oblate, only 3 chambers visible externally, low trochospiral
coiling. Chambers initially spherical, rapidly enlarging, becoming subspherical and
heavily covered by deposition of secondary cortex. Small abortive chambers (not bullae) occasionally occur. Apertures sutural, primary aperture interiomarginal, umbilical;
secondary apertures also sutural. Chambers modified at apertures by thick, imperforate
flange-like lips. Wall calcareous, thick, perforated more coarsely on portions of chambers away from flanges, pores set in shallow depressions and have small upraised collars,
non-spinose in the adult stage, may be spinose similar to *G. sacculifer* in juveniles.

Remarks: In its gross morphology, this genus is monotypic, discounting the variety

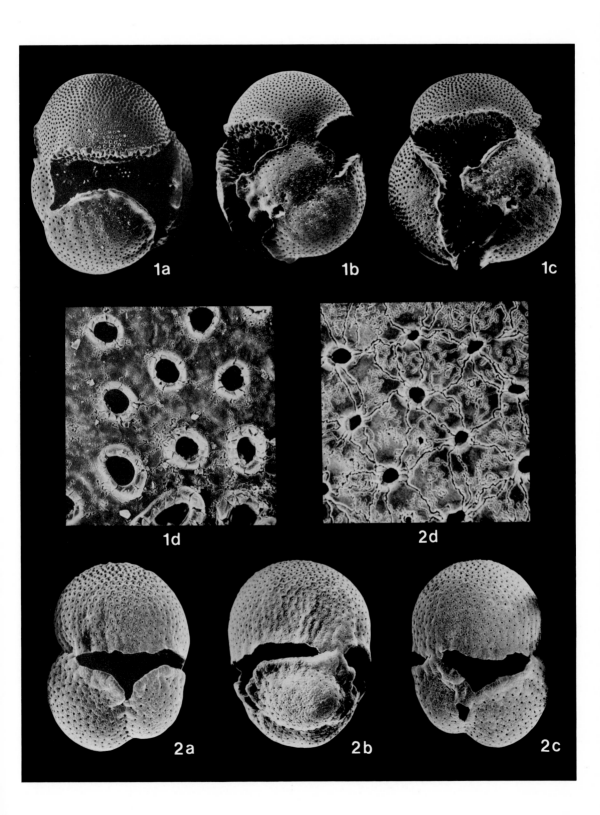

excavata. Bé (1965) and BANDY, INGLE and FRERICHS (1967), however, have suggested by dissection the possibility that *dehiscens* may be a deep-water stage of *G. sacculifer*. On the other hand, *dehiscens* has also been shown to evolve separately from the *Sphaeroidinellopsis* group which appeared in the Early Miocene (BANNER and BLOW, 1965, p. 1164).

Distribution: Early Pliocene (N. 19) to Recent, equatorial to temperate waters.

21 *Candeina nitida* D'ORBIGNY, 1839
Plate 21, Figs. 1a–c, 2a–c, × 93; 1d, 2d, × 1285

Candeina nitida D'ORBIGNY, 1839a, p. 108. pl. 2, figs. 27–28. — BOLLI, LOEBLICH and TAPPAN, 1957, p. 35, pl. 6, figs. 10a–c, 11.
Candeina nitida nitida D'ORBIGNY. — BLOW, 1969, pp. 335, 384–386, pl. 23, figs. 1–4. — BRÖNNIMANN and RESIG, 1971, pp. 1290–1291, pl. 14, figs. 7–8.
Candeina nitida D'ORBIGNY var. *triloba* CUSHMAN, 1921, p. 296, pl. 57, figs. 1a–c.
Candeina milletti DOLLFUS, 1905, pp. 692–693, pl. 7, figs. 2a–c.

Types: Holotype (*nitida*)—from the marine sands of Cuba and Jamaica.
Plesiotype (*nitida*; Figs. 1a–d)—V3-3, core top (Lat. 18°51′N, Long. 67°07′W, 2,661 m).
Holotype (*milletti*)—found only in Area 2 from Singapore in the north to Banka in the south, Sumatra in the west and Borneo in the east, Malay Archipelago; very shallow water.
Plesiotype (*milletti*; Figs. 2a–d)—V28-222, trigger core top (Lat. 11°19′ S, Long. 174°32′ E, 2,933 m).

Diagnosis: Test size medium to large, a low to medium height trochospiral coil, 3–4 globular chambers in the final whorl, about 15 chambers in all arranged in about 3 whorls. Chambers spherical, increasing rapidly in size, greatly embracing, sutures distinct. Primary aperture umbilical, a low arch with a distinct rim-like lip; secondary apertures sutural, a continuous series of rounded openings along both intercameral and spiral sutures, each with a round arched lip. Wall calcareous, very finely perforated by tiny, irregularly shaped pores, non-spinose.

Remarks: While the morphology of this genus renders it distinctly easy to identify, workers have had trouble assigning it to a phylogenetic position. Careful dissection and SEM studies by SAITO and THOMPSON (in preparation) have determined its phylogenetic link with *Globigerinita glutinata* which can be traced to a common ancestor (referred to as *Globigerinoides parkerae* BERMÚDEZ by BLOW, 1969, p. 384, pl. 22, figs. 1–4 and *Candeina nitida praenitida* BLOW, 1969, pl. 27, figs. 5–8) in the Middle Miocene (N. 16). Concerning the phylogenetic position of this species, it is of particular interest to note that the earlier-formed whorls often have an umbilical aperture similar to that of *G. glutinata* and that the characteristic sutural apertures with their raised "collars" are restricted to the final whorl. It should also be noted that the "collars" of the apertures usually extend across the groove of the spiral suture to the preceding chamber, forming a complete ring; where this aperture intersects an intercameral suture as well, the opening enlarges along the triple-junction. BRÖNNIMANN and RESIG (1971, pl. 14, fig. 7) illustrated a specimen showing the development of a "secondary crust" with very fine pores covering up much larger pores on the earlier crust of the same chamber. The form *milletti* is believed to be an aberrant morphotype of *C. nitida* differing in the low height of the trochospire. CUSHMAN's *C. nitida triloba* shows a complete envelopment of the earlier whorls by the last three chambers.

Distribution: Late Miocene (N. 17) to Recent, equatorial and tropical waters.

Plate 21   *Candeina nitida* D'ORBIGNY, 1839

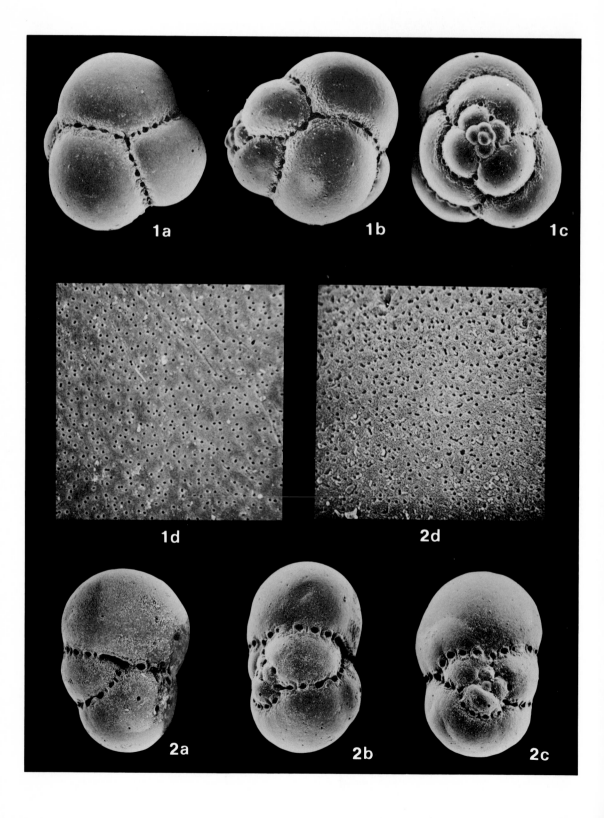

22   *Globigerinita glutinata* (EGGER), 1893
Plate 22, Figs. 1a–c, 2–7, × 133; 1d, × 1666
(see also Pl. 23, Figs. 1a–c)

Globigerina glutinata EGGER, 1893, p. 371, pl. 13, figs. 19–21.
Tinophodella ambitacrena LOEBLICH and TAPPAN, 1957, p. 114, figs. 2,3.
Globigerinita naparimaensis BRÖNNIMANN. — BOLLI, LOEBLICH and TAPPAN, 1957, p. 37, pl. 8, figs. 2a–c.
Globigerinoides parkerae BERMÚDEZ, 1961, p. 1232, pl. 10, figs. 10, 11. — BLOW, 1969, p. 325, pl. 22, figs. 1–4.
Globigerinita glutinata (EGGER). — PARKER, 1962, pp. 246–249, pl. 19, figs. 1–16. — PARKER, 1967, p. 146, pl. 17, figs. 3–5. — HERB, 1968, pp. 478–479, pls. 3, figs. 4–6.
Globigerinita glutinata (EGGER) *flparkerae* BRÖNNIMANN and RESIG, 1971, pp. 1303–1306, pl. 23, figs. 1–4; pl. 50, fig. 6, text-figs. 15a–d.
Globigerinita glutinata glutinata (EGGER). — BRÖNNIMANN and RESIG, 1971, p. 1306, pl. 23, fig. 5.
Globigerinita glutinata ambitacrena (LOEBLICH and TAPPAN). — FLEISHER, 1974a, p. 1022, pl. 9, fig. 3.
Globigerinita glutinata parkerae (BERMÚDEZ). — FLEISHER, 1974a, p. 1022.

Types: Syntypes (*G. glutinata*)—Station 17 (Lat. 10° 12.9′ N, Long. 17° 25.5′ W, 667 m); Station 90 (Lat. 18° 52′ S, Long. 116° 18′ E, 359 m); Station 101A (near Amboina Is. 55 m), Station 102 (Lat.02° 54.5′ S, Long.127° 46.5′ E, 3,145 m).

Plesiotype (*G. glutinata*)—V24-150, trigger core top (Lat. 02° 12′ S, Long. 155° 42′ E, 1,849 m).

Holotype (*G. parkerae*)—Canal de Micolas, north of Cuba (Lat. 23° 15′ N, Long. 79° 57′ W, 380 Borzas).

Plesiotype (*G. parkerae*)—V28-238, trigger core top (Lat. 01° 01′ N, Long. 160° 29′ E, 3,120 m)

Diagnosis: Test small to medium in size, a medium height trochospire, 4 globular chambers in the final whorl, about 10 chambers arranged in 2 $1/2$ whorls. Chambers spherical to slightly flattened radially, increasing moderately in size as added, partially embracing, sutures distinct; bullae may appear along sutures on any part of the test. Primary aperture umbilical, a low but wide symmetrical arch with a thin rim-like lip. Secondary apertures also with lips may be developed on the spiral side of the test at the intersection of the spiral and intercameral suture. Wall calcareous, fragile, smooth to finely hispid and non-spinose, irregularly perforated with fine pores.

Remarks: As illustrated in the accompanying plate, the bulla variability gives this species many morphologic variations even within a single sample, ranging from *G. glutinata* EGGER to *T. ambitacrena* LOEBLICH and TAPPAN. It is also possible that a fully inflated bulla with infralaminal apertures, and not a true chamber, gives the form called *G. parkerae* (= *G. flparkerae*) what are often interpreted as secondary apertures. A well lobulated form identified as *Globigerina juvenilis* BOLLI described from Miocene sediments is often recorded from Recent sediments.

Distribution: Middle Miocene (N. 13) to Recent, cosmopolitan, although the forms with supplementary apertures are not as abundant as the normal form with one aperture.

Plate 22 *Globigerinita glutinata* (EGGER), 1893

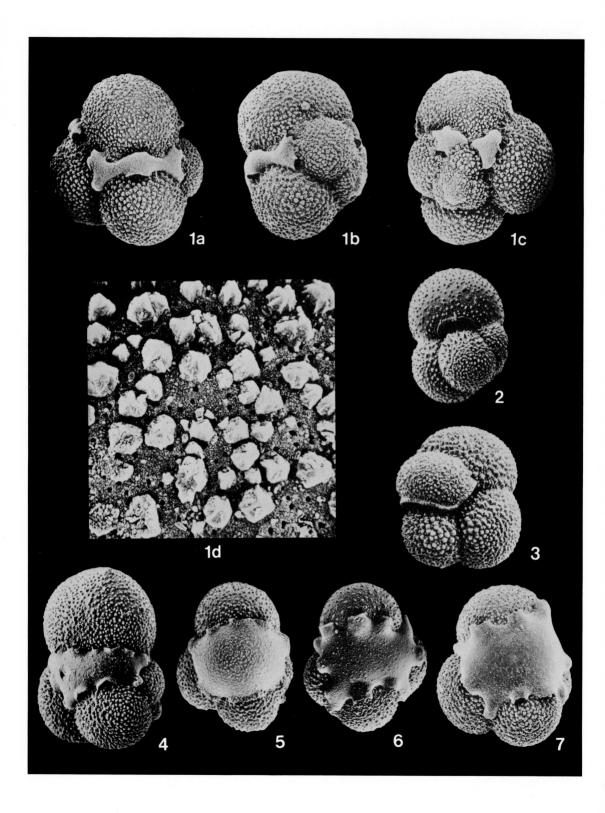

23. 1 *Globigerinita glutinata* (EGGER), 1897—variant form
Plate 23, Figs. 1a–c, × 240; 1d, × 1200

Globigerinoides parkerae BERMÚDEZ, 1961, p. 1232, pl. 10, figs. 10, 11. — BLOW, 1969, p. 325, pl. 27, figs. 1–4.

Globigerinita glutinata (EGGER) *flparkerae* BRÖNNIMANN and RESIG, 1971, pp. 1303–1306, pl. 23, figs. 1–4; pl. 50, fig. 6; text-fig. 15.

Types: Plesiotype—V28-238, trigger core top (Lat. 01°01′N, Long. 160°29′E, 3,120 m).

Remarks: This form is essentially similar to *G. glutinata,* and here is considered a junior synonym. In the past, it has been isolated as a separate species because of its tendency to form one or more (*sic*) supplementary apertures, each having a lip. Bearing in mind, however, the range of variation of the bullae of this species, it is also possible that what seems to be the final chamber is rather a large bulla.

Distribution: Middle Miocene (N. 13) to Recent, similar habitat to typical *G. glutinata.*

23. 2 *Globigerinita iota* PARKER, 1962
Plate 23, Figs. 2a–c, × 240; 2d, × 1200

?Globigerina radians EGGER. — RHUMBLER, 1911, p. 148, pl. 29, figs. 2–4.

Globigerina sp. PARKER, 1954, p. 476.

Globigerinita iota PARKER, 1962, p. 250, pl. 10, figs. 26–30.

Turborotalita iota PARKER. — LIPPS, 1966, p. 1267. — BRÖNNIMANN and RESIG, 1971, p. 1324, pl. 23, figs. 6,7,8. — RÖGL and BOLLI, 1973, p. 571, pl. 8, figs. 17–23; pl. 16, fig. 9.

Types: Holotype—Downwind Station BG 73, 0–2 cm (Lat. 43°48′S, Long. 108°09′W, 3,080 m). LIPPS, 1966, p. 1267.

Topotype—Same core, 2–4 cm.

Diagnosis: Test small, a low trochospire, 4–5 globular chambers in the final whorl, about 13 chambers in all arranged in 2 $1/2$–3 whorls. Chambers subspherical to slightly flattened radially, partially embracing, sutures distinct. Aperture umbilical-extraumbilical with a tongue-like apertural flap, but typically covered by a slightly inflated bulla with infralaminal apertures along intercameral sutures of the final whorl. Both primary aperture and infralaminal apertures frequently have a thin lip. Wall calcareous, very finely perforated, smooth to finely granular surface, non-spinose.

Remarks: This species is more compressed and has a lower trochospire than *G. glutinata,* and differs from *T. humilis* by its infralaminal apertures. PARKER (1962, pl. 10, fig. 30) also included in the concept of this species a form typically lacking bullae but was otherwise hard to differentiate from those possessing a bulla. BRÖNNIMANN and RESIG (1971) regarded the infralaminal bulla as an "umbilical extension of the apertural walls," and transferred it to *Turborotalia.* Here, however, the structure is considered a bulla.

Distribution: Pleistocene to Recent. PARKER (1962) reports it to be cosmopolitan to at least 60°S. latitude in the Pacific. ,

Plate 23   *Globigerinita glutinata* (EGGER), 1897—variant form
Globigerinita iota PARKER, 1962

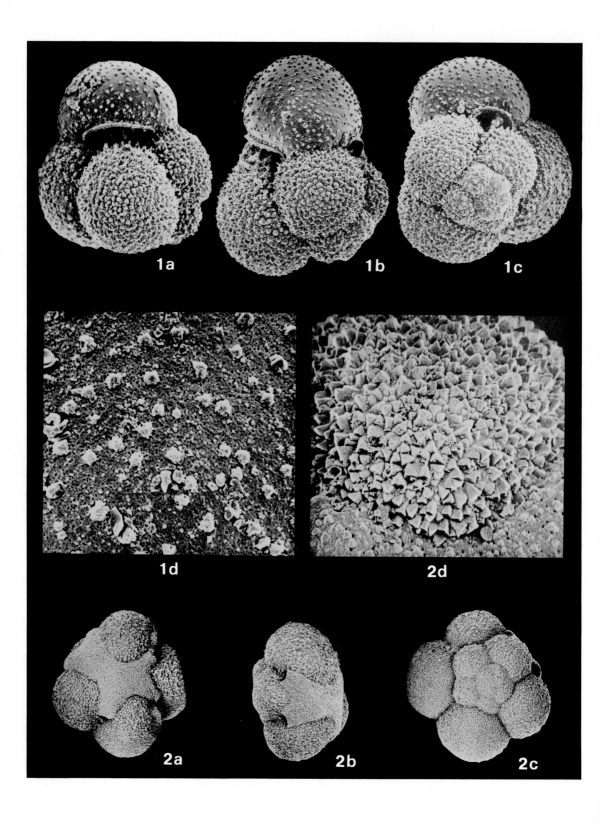

24. 1 *Globigerinita minuta* (NATLAND), 1933
Plate 24, Figs. 1a–b,2, × 400; 1c, × 4000

Globigerinoides minuta NATLAND, 1933, line 34 of table (*nomen nudum*). — NATLAND, 1938, p. 150, pl. 7, figs. 2,3.
Globigerinoides cf. *minuta* NATLAND. — BRADSHAW, 1959, p. 40, pl. 7, figs. 9–11.
Globigerinita uvula minuta (NATLAND). — RÖGL and BOLLI, 1973, pp. 571–572, pl. 5, fig. 25; pl. 15, fig. 8.

Types: Holotype—Locality 117 (Lat. 32°22′16″N, Long. 118°18′40″W, 122 m).
Plesiotype—V20-119, core top (Lat.47°57′N, Long.168°47′E, 2,739 m).

Diagnosis: Test small, a medium to high trochospire, 3–4 globular chambers in the final whorl, about 12 chambers in all arranged in about 3 whorls. Chambers spherical, increasing moderately in size as added, partially embracing, sutures distinct. Aperture umbilical, typically covered by a bulla-like final chamber often with an additional (*sic*) supplementary aperture. Wall calcareous, very finely and irregularly perforated, smooth to slightly granular, non-spinose.

Remarks: Although PARKER (1962) considered *G. minuta* to be a warm water phenotype of *G. uvula,* forming a gradational series, the two forms can be distinguished and are treated separately here. *G. minuta* has more inflated chambers, a lower trochospire and is more granular than *G. uvula.*

Distribution: NATLAND (1938) reports it from Pliocene and Pleistocene sediments of California. Cosmopolitan, according to BRADSHAW (1959), although preferring warmer waters than *G. uvula.*

24. 2 *Globigerinita uvula* (EHRENBERG), 1861
Plate 24, Figs. 3a–c, × 400; 3d, × 4000

Pylodexia uvula EHRENBERG, 1861, pp. 276,277,308. — EHRENBERG, 1873, pl. 2, figs. 24–25.
Globigerina sp. BRADY, 1884, p. 603, pl. 82, figs. 8–9.
Globigerina bradyi WIESNER, 1931, pp. 133–134 (no figures). — BANNER and BLOW, 1960a, p. 5, pl. 3, figs. 1 (lectotype), 2.
Globigerina elevata D'ORBIGNY. — HERON-ALLEN and EARLAND, 1932 (not D'ORBIGNY, 1840), p. 402.
Globigerinita uvula (EHRENBERG). — PARKER, 1962, p. 252, pl. 8, figs. 14–26. — PARKER, 1967, p. 146. pl. 17, figs. 8–9. — HERB, 1968, p. 479, pl. 3, figs. 7–8.
Globigerinita uvula uvula (EHRENBERG) — RÖGL and BOLLI, 1973, p. 571, pl. 5, fig. 24; pl. 15, fig. 7.
Globigerinoides elongata bikiniensis McCULLOCH, 1979, p. 418, pl. 174, fig. 2.

Types: Holotype (*uvula*)—Greenland Deep, North Atlantic (Lat. 62°40′N, 29°W, 6,000 m).
Holotype (*bradyi*)—Station 39 (Lat. 46°47′S, Long. 50°37′E, 2,320 m).
Lectotype (*bradyi*)—Challenger Station 144.
Plesiotype (*uvula*)—V16-125, trigger core top (Lat. 47°01′S, Long. 179°15′W, 2,953 m).

Diagnosis: Test very small, a high trochospire, 3–4 globular chambers in the final whorl, 15–20 chambers in all arranged in 3–4 whorls. Chambers spherical, increasing slowly in size as added, much embracing, sutures distinct. Aperture interiomarginal, umbilical, a small, low arch with no apparent modification. Wall calcareous, fragile, translucent, very smooth and very finely perforated, non-spinose but slightly pustulate near the aperture.

Remarks: This very small but distinctive species is usually found in the $<149\mu$ size fraction of foraminiferal samples from high latitudes. It can be distinguished from *G. minuta* by its higher trochospire and translucent, smooth test.

Distribution: Pleistocene(?) to Recent, high latitude assemblages.

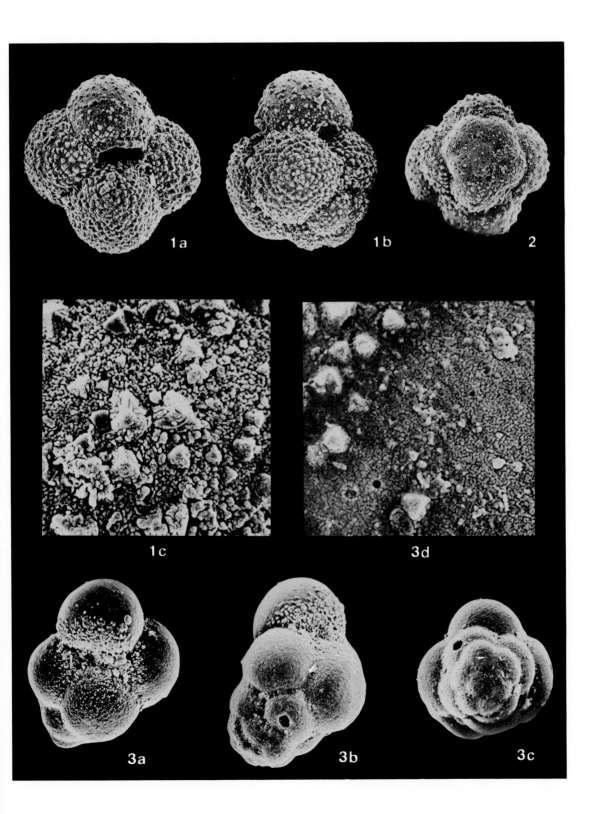

25 *Turborotalita humilis* (BRADY), 1884
Plate 25, Figs. 1a–b, 2, × 280; 3a–b, 4, × 400; 1c, 3c, × 2400

Truncatulina humilis BRADY, 1884, p. 665, pl. 94, fig. 7 — BANNER and BLOW, 1960a, p. 36, pl. 8, fig. 1 (lectotype).

Globigerina lamellosa TERQUEM — RHUMBLER (not TERQUEM, 1882), 1911, pl. 30, figs. 1–6.

Globigerina cretacea D'ORBIGNY var. *eggeri* HERON-ALLEN and EARLAD, 1922, p. 188, pl. 7, figs. 6–8.

Globigerina cristata HERON-ALLEN and EARLAND, 1929, p. 331, pl. 4, figs. 33–39. — BANNER and BLOW, 1960a, p. 10, pl. 7, fig. 5 (lectotype).

Valvulerina cf. *humilis* (BRADY). — PHLEGER and PARKER, 1951, p. 25, pl. 13, figs. 9–10.

Globigerina sp. PARKER, 1954, pp. 476–477.

Globigerinita parkerae LOEBLICH and TAPPAN, 1957, p. 113, fig. 1.

Globigerinita humilis (BRADY). — PARKER, 1962, p. 249, pl. 10, figs. 1–25. — CIFELLI and SMITH, 1970, p. 36, pl. 5, fig. 1.

Turborotalita humilis (BRADY). — BLOW and BANNER, 1962, p. 122. — LIPPS, 1966, p. 1267. — PARKER, 1967, pp. 146–147, pl. 17, fig. 10. — BRÖNNIMANN and RESIG, 1971, p. 1324, pl. 21, figs. 6,7.

Globigerinita cretacea saratogaensis BOLTOVSKOY (not *Globigerina cretacea* var. *saratogaensis* Applin, 1925), 1966, pp. 13–14, pl. 1, figs. 12a–c.

Globigerinita bikiniensis McCULLOCH, 1979, p. 419, pl. 175, figs. 13,14.

Globigerina (?) *cristatiformis* McCULLOCH, 1979, p. 412, pl. 175, figs. 12.

Globigerina (?) *guadalupensis* McCULLOCH, 1979, p. 413, pl. 175, figs. 9–10.

Turborotalita cristata (HERON-ALLEN and EARLAND). — IACCARINO and SALVATORINI, 1979, pl. 7, figs. 30, 31.

Types: Syntype (*humilis*)—Challenger Station 5 (Lat. 24°20′N, Long. 24°28′W, 2,740 m).

Lectotype (*humilis*)—from remaining original materials.

Near-Topotype(*humilis*; Figs.1a–c,2)—V10-84, trigger core top (Lat. 24°23.5′N, Long. 24°03.5′W, 5,255 m).

Holotype (*cristata*)—Discovery Station (Lat. 26°17′40″S, Long. 14°26′28″E, 3,170 m).

Near-Topotype(*cristata*; Figs.3a–c,4)—RC8-93, trigger core top (Lat. 29°22′S, Long. 105°14′W, 3,157 m).

Holotype (*parkerae*)—Lat. 29°04′N, Long. 85°49′W, 189 m.

Diagnosis: Test small in size, about 5–7 globular chambers in the final whorl, about 16 chambers in all arranged in 2 ½ very low trochospiral whorls. Chambers subglobular to ovate radially, considerably embracing, sutures well incised; final chamber extends basally to cover the umbilicus like a bulla, and has numerous small, arched infralaminal apertures. Aperture (except when covered by bulla) interiomarginal, umbilical. Wall calcareous, thin, finely and irregularly perforated except at the bulla, nearly imperforate, hispid and possibly spinose. Distal ends of chambers often coarsely hispid.

Remarks: This species or group can be distinguished from *G. quinqueloba* by both its smaller size and more numerous chambers. CIFELLI and SMITH (1970) figured the first specimens recorded in plankton tows. Considerable variation in form is represented by this species, which PARKER (1962) attributed to ecological variations. The form described as *G. cristata* is considerably more lobulate than typical *T. humilis* and the specimen illustrated by CIFELLI and SMITH (1970) has a much less-developed bulla than most illustrations.

Distribution: PARKER (1962) reported *T. humilis* as widely distributed in the South

Plate 25   *Turborotalita humilis* (BRADY), 1884   85

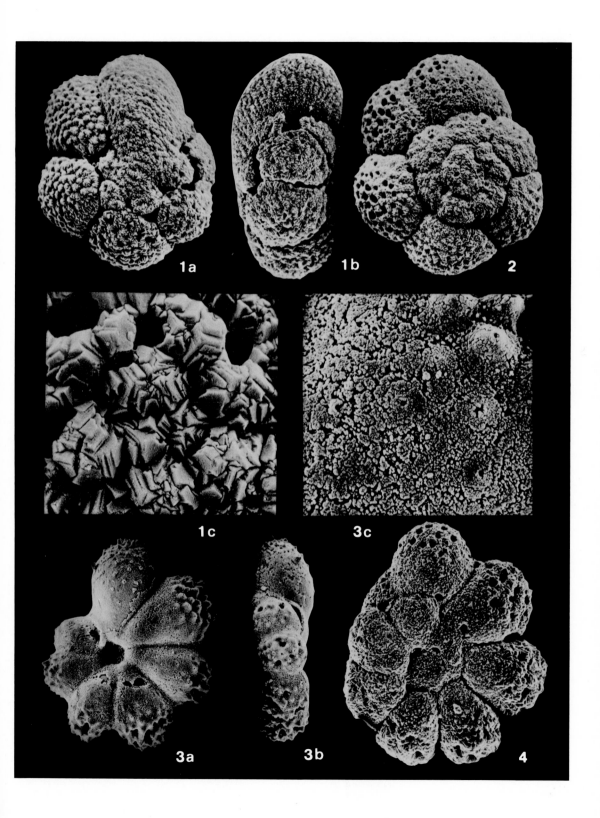

Pacific and CIFELLI and SMITH (1970) found some in the northwest Atlantic. *G. parkerae* came from the Gulf of Mexico and *G. cristata* came from the southeast Atlantic. Many of the specimens illustrated by HERON-ALLEN and EARLAND are particularly hispid on the distal ends of the chambers, which is not distinct in small (juvenile) specimens.

26. 1 *Berggrenia pumilio* (PARKER), 1962
Plate 26, Figs. 1a–b, 2, × 392; 1c, × 2143

Globorotalia pumilio PARKER, 1962, p. 238, pl. 6, figs. 2,3 — PARKER, 1976, p. 259, pl. 1, figs. 3–5.
Eoglobigerina pumilio (PARKER) — LIPPS, 1966, p. 1266.
Globanomalina pumilio (PARKER) — PARKER, 1967, pp. 148–149, pl. 18, fig. 5.
Turborotalita pumilio (PARKER). — FLEISHER, 1974a, p. 87, pl. 21, fig. 7.

Types: Holotype—Capricorn HG 41, 0–1 cm (Lat. 15°55.5′S, Long. 117°13.8′W, 3,394 m).

Topotype—Capricorn HG 41, 1–3 cm.

Diagnosis: Test very small in size, 5 $1/2$–7 globular chambers in the final whorl, up to 2 chambers in all arranged in 2 $1/2$–3 whorls. Chambers subspherical, partially embracing, increasing slowly in size as added, sutures very slightly curving. Aperture umbilical-extraumbilical, a low arch often with a thin rim-like lip. Wall calcareous, thin, irregularly perforated, mostly on the spiral side near the sutures, does not apear to be spinose.

Remarks: Although a frequent component of South Pacific sediments, it is seldom recorded due to its small size. Its morphology is distinctive, and only slightly resembles its ancestral form *B. praepumilio,* from which it differs by having more chambers, more restricted pore locations and more trochoid coiling.

Distribution: PARKER (1967) reports it from the South Pacific north of 45°S in Quaternary sediments.

26. 2 *Berggrenia clarkei* (RÖGL and BOLLI), 1973
Plate 26, Figs. 3a–c, × 392; 4, × 2143

Globigerina aff. *Globigerina quinqueloba* NATLAND. — BRÖNNIMANN and RESIG, 1971, p. 1300, pl. 43, figs. 8–9.
Globigerina clarkei RÖGL and BOLLI, 1973, p. 563, pl. 14, figs. 13–15.
Turborotalita "quinqueloba" (NATLAND). — FLEISHER, 1974b, p. 994, pl. 2, figs. 4–6.

Types: Holotype—DSDP Leg. 15, Hole 147, Core 4, core catcher, 32 m (Lat. 10° 42.48′N, Long. 65°10.48′W).

Near-Topotype—V12-98, 0–10 cm (Lat. 10°47.1′N, Long. 65°06.8′W, 736 m).

Diagnosis: Test small in size, 5 globular chambers in the final whorl, 9–13 chambers in all arranged in about 2–2 $1/2$ low trochospiral whorls. Chambers subspherical, much embracing, final chamber slightly elongate radially, chamber size increases slowly as added. Sutures distinct, slightly incised. Aperture umbilical-extraumbilical, a low slit shaped by a rim-like lip on the base of the final chamber. Wall calcareous, thick, irregularly perforated, very smooth, non-spinose. Some specimens are slightly hispid.

Remarks: The basic shape of the test is very reminiscent of *T. quinqueloba,* differing in its smaller size and large, irregulary-spaced pores. Many of these pores occur along the spiral suture on the dorsal side, a feature possessed by *B. pumilio* and *B. praepumilio.* It differs from *T. humilis* having fewer chambers and lacks the bulla-like final chamber of *T. humilis.*

Distribution: RÖGL and BOLLI (1973) recorded it from Late Pleistocene to Recent in the Caribbean; BRÖNNIMANN and RESIG (1971) recorded tubular form from Middle Pliocene to Middle Pleistocene in the western equatorial Pacific.

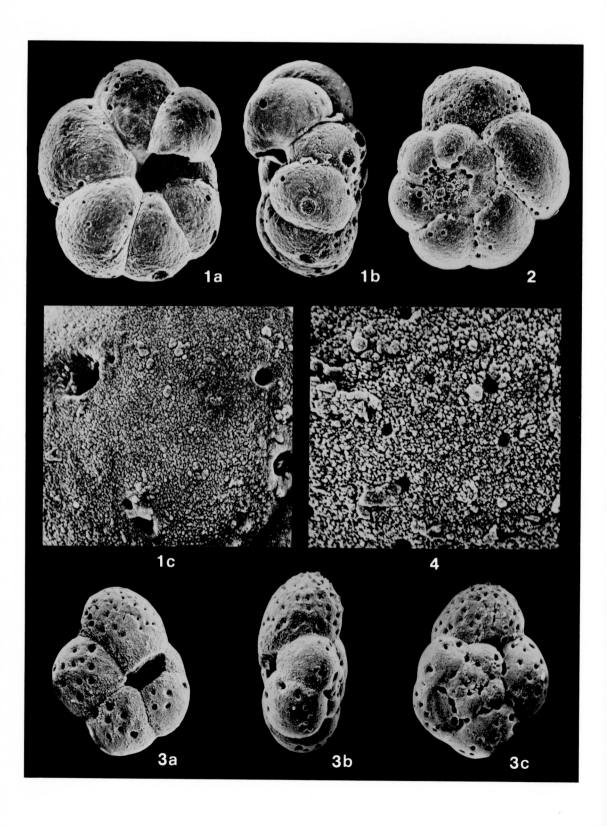

27. 1 ?*Globorotalia pseudopumilio* Brönnimann and Resig, 1971
Plate 27, Figs. 1–3, × 392; 4, × 2143

Globorotalia (Turborotalia) pseudopumilio Brönnimann and Resig, 1971, pp. 1282–1283, pl. 43, figs. 4,5.

Types: Holotype—DSDP Hole 62.1, Core 6, Section 5, 15–17 cm (Lat. 01°52.2′N, Long. 141°56.3′E, 2,607 m).

Plesiotype—V28-239, 2,099 cm (Lat. 03°15′N, Long. 159°11′E, 3,490 m).

Diagnosis: Test small in size, 5 globular chambers in the final whorl, about 12 chambers in all arranged in 2 $1/2$ low trochospiral whorls. Chambers subspherical, partially embracing, increasing slowly in size as added; sutures distinct but shallow and wide. Aperture umbilical-extraumbilical, a circular arch flanked by a low lip. Wall calcareous, uniformly perforated, pores surrounded by wall-like structures formed by coalescing crystals, some of which may possess spine base remnants.

Remarks: *G. pseudopumilio* is closely comparable with *G. pumilio,* but is slightly larger and has 5 instead of 6 or more chambers in the final whorl and is more lobulate. Of particular interest are the wide and funnel-shaped pore depressions in *G. pseudopumilio,* formed by merging pustules. This gives the test a distinctive honeycombed surface appearance. It is unfortunate that Brönnimann and Resig (1971) only figured two spiral views of this curious new taxon.

Distribution: Recorded only from Early Pliocene (N. 19) to Early Pleistocene (N. 22) from the western equatorial Pacific.

27. 2 *Berggrenia praepumilio* (Parker), 1967
Plate 27, Figs. 5a–b,6,× 392; 5c, × 2143

Globanomalina praepumilio Parker, 1967, p. 148, pl. 18, figs. 1–4.
Berggrenia praepumilio (Parker). — Parker, 1976, p. 259, pl. 1, figs. 1,2; text-figs. 1–2.

Types: Holotype—Dodo 117 P, 164–166 cm (Lat. 18°21′S, Long. 62°04′E, 3,398 m).

Topotype—Dodo 117 P, 170 cm.

Diagnosis: Test small in size, 5 $1/2$–8 globular chambers in the final whorl, about 15 chambers in all arranged in a very low trochospire often nearly planispiral of about 2 $1/2$ whorls. Chambers subspherical, increasing very slowly in size as added, partially embracing, sutures slightly curved. Aperture interiomarginal, umbilical-extraumbilical, a low slightly asymmetrical arch with a narrow rim-like lip. Wall calcareous, thin, very finely perforated, mostly on the spiral side, possibly non-spinose.

Remarks: Although quite small and difficult to handle, this species is quite distinctive in its nearly planispiral appearance under the light microscope. This "planispirality" is due, however, to a depression of the earlier whorls by the later, more inflated chamber, and the location of the aperture along the base of the final chamber. It is noteworthy that the spiral side is more porous than the umbilical side, with some of the pores lying along the spiral suture.

Distribution: Upper Pliocene (N. 21) to Pleistocene? (N. 22), Pacific.

Plate 27   *?Globorotalia pseudopumilio* Brönnimann and Resig, 1971
Berggrenia praepumilio (Parker), 1967

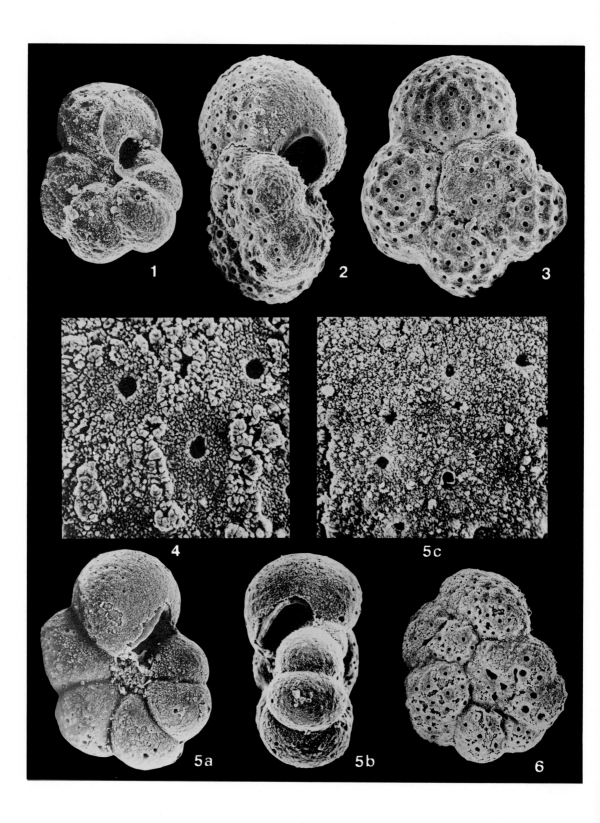

28   *Globorotaloides hexagona* (NATLAND), 1938
Plate 28, Figs. 1a–c, × 116; 1d, × 1166; 2a–c, × 116; 2d, × 1166

Globigerina hexagona NATLAND, 1938, p. 149, pl. 7, figs. 1a–c.
"*Globigerina*" *hexagona* NATLAND. — BRÖNNIMANN and RESIG, 1971, p. 1295, pl. 2, figs. 1–5.
Globoquadrina hexagona (NATLAND). — PARKER, 1962, p. 244, pl. 8, figs. 5–13.
Globigerina (Globorotaloides) hexagona NATLAND. — TODD, 1964, pp. 1080–1081, pl. 202, fig. 3.
Globorotalia (Clavatorella) oveyi BUCKLEY, 1973, p. 169, pl. 1, figs. 1–10.
Globorotaloides hexagonus (NATLAND). — FLEISHER, 1974a, p. 1028, pl. 13, fig. 6.
Globorotaloides hexagona (NATLAND). — SAITO, THOMPSON and BREGER, 1976, p. 286, pl. 4, fig. 3; pl. 7,
    fig. 3; pl. 8, fig.6.
Globigerina clippertonensis MCCULLOCH, 1979, p. 411, pl. 173, figs. 1,2.

Types: Holotype (*hexagona*)—Locality 111 (Lat. 33°27′20″N, Long. 118°19′00″W,
    884 m).

   Plesiotype (*hexagona*; Figs. 1a–d)—RC10-80 (Lat. 11°54.3′N, Long. 109°
    16.9′W, 3,288 m).

   Holotype (*oveyi*; Figs. 2a–d)—R/V Anton Bruun I, May 1963, Sta. 97,
    Smithsonian Sorting Center, no. 479 (Lat. 13°08′N, Long. 86°12′E, 0–250 m
    plankton tow).

   Topotype (*oveyi*)—Sample no. 479 (specimen broken in handling).

Diagnosis: Test medium to large, very low trochospiral coil, 5–6 chambers in the final
whorl, about 12 globular chambers in about 2 1/2 whorls. Chambers spherical becoming
slightly radially elongate, partially embracing, intercameral sutures nearly radial to
slightly curved, deeply incised. Aperture umbilical to umbilical-extraumbilical, a low
asymmetrical arch typically obscured from umbilical view by a thick and broad apertural flap on each chamber slightly attached between chambers. Wall calcareous, very
coarsely perforated with large deep pores set in hexagonal or pentagonal funnel-like
depressions with practically no interpore area. Pustulate only near sutures or in the umbilicus.

Remarks: The very cancellate surface with its large pores is the most characteristic
feature of this species, distinguishing it from *Globoquadrina*. Chamber size and shape
vary greatly in this species, and in the Indian Ocean led BUCKLEY to create a new species
of *Clavatorella*. As shown by SAITO, THOMPSON and BREGER (1976), however, *Clavatorella* is less cancellate and does not possess the pustules of *G. hexagona*. BRÖNNIMANN's
reexamination of the holotype (BRÖNNIMANN and RESIG, 1971) and additional observation of the *G. hexagona* group seems to point out the difficulties of trying to define
a species (or genus) on the basis of apertural position alone.

Distribution: Latest Miocene (N. 18–N. 19) to Recent, subtropical waters. PHLEGER
et al. (1953) and ERICSON *et al.* (1961) recorded its extinction in the Atlantic in the Late
Pleistocene, although it is still present in plankton and sediments of the Indo-Pacific.

Plate 28 *Globorotaloides hexagona* (NATLAND), 1938

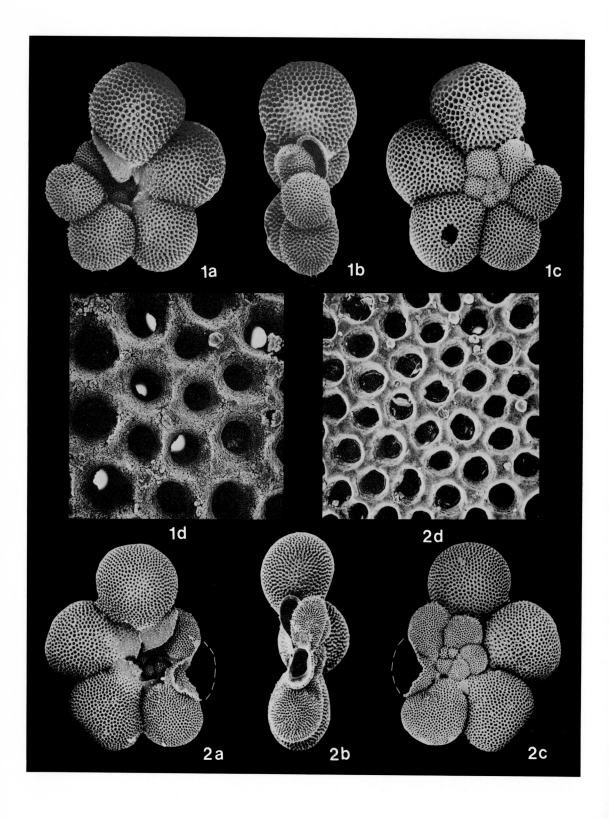

1a

1b

1c

1d

2d

2a

2b

2c

29. 1 *Globoquadrina pseudofoliata* PARKER, 1967
Plate 29, Figs. 1a–c, × 71; 1d, × 714

Globoquadrina pseudofoliata PARKER, 1967, pp. 170–171, pl. 27, figs. 1–3.
Globigerina pseudofoliata (PARKER) — BRÖNNIMANN and RESIG, 1971, p. 1300, pl. 4, figs. 1–3,5,6,8–10; text-figs. 11, 12.

Types: Holotype—LSDH 78P, 30–32 cm (Lat.04°31′S, Long.168°02′E, 3,208 m).
Topotype—LSDH 78P, 506–508 cm.

Diagnosis: Test large, low trochospire, 4 globular chambers in the final whorl, about 10–15 chambers in all arranged in about 2 $1/2$ whorls. Chambers spherical to subspherical, rapidly enlarging as added, partly embracing with deep sutures. The final chamber overhangs the umbilicus. Early sutures distinctly curved, later becoming more straight. Aperture umbilical, usually partly to almost completely blocked by the final chamber, which has a large and wide imperforated umbilical tooth. Wall calcareous, coarsely perforated with large pores of uniform diameter and spacing set in polygonal, funnel-shaped pore depressions, non-spinose but finely pustulate, usually at the intersections of several interpore ridges.

Remarks: In equatorial Indo-Pacific sediments, this typical form can be distinguished from *G. hexagona* by its more inflated chambers and less-cancellate surface, and from *G. conglomerata* by its lower trochospire, more lobulate periphery and early curved sutures. Near the edges of its geographic range, these distinctions become more difficult since the chambers of *G. pseudofoliata* do not inflate or separate fully. Here, the initially curved sutures serve to distinguish even small atypical specimens from *G. conglomerata*. BRÖNNIMANN and RESIG (1971) illustrated specimens representing several stages of development, noting that the juveniles were often nearly imperforate (pl. 4, fig.2) and do not have apertural flaps.

Distribution: Early Pliocene (N. 19) to latest Pleistocene, Indo-Pacific, 10°N–20°S. THOMPSON and SAITO (1974) date the extinction level of this species at 220,000 years B.P.

29. 2 *Globoquadrina conglomerata* (SCHWAGER), 1866
Plate 29, Figs. 2a–c, × 71; 3, × 714

Globigerina conglomerata SCHWAGER, 1866, pp. 255 – 256, pl. 7, fig. 113. — BANNER and BLOW, 1960a, p. 7, pl. 2, fig. 3 (neotype). — BRÖNNIMANN and RESIG, 1971, pp. 1292–1293, pl. 5, fig. 8; text-fig. 6.
Globoquadrina conglomerata (SCHWAGER). — PARKER, 1962, p. 240, pl. 6, figs. 11–18.

Types: Holotype—Neogene deposits on Kar Nikobar, Andaman Islands, India.
Neotype—From metatypes sent to BRADY by SCHWAGER; BANNER and BLOW, 1960a, p. 7.
Plesiotype—Core RC11-210, 514cm (Lat. 01°49′N, Long. 140°03′W, 4,420 m).

Diagnosis: Test large, low to medium height trochospire, moderately lobulate, 3 globular chambers in the final whorl, about 10 chambers arranged in about 2–2 $1/2$ whorls. Chambers initially spherical or subspherical becoming flattened radially, size increasing moderately as added, considerable overlap with moderate suture incisement. Aperture umbilical, a large rounded-triangular or quadrate opening occasionally showing an umbilical tooth. Wall calcareous, coarsely perforated with large circular pores of uniform spacing, each in a polygonal pore depression, non-spinose but coarsely pustulate, particularly in the vicinity of the large umbilicus.

Remarks: *G. conglomerata* is larger in overall size and more inflated than *G. hexagona,* but less lobulate than *G. pseudofoliata. Globoquadrina venezuelana* (Hedberg) is the ancestral form of both *G. conglomerata* and *G. pseudofoliata,* is less lobulate with more embracing chambers than *G. conglomerata,* and became extinct in the Early Pliocene.

Distribution: Early Pliocene (N. 19) to Recent, equatorial to subtropical waters. PARKER (1965) notes that it became extinct in the Atlantic in the early Pleistocene, but is flourishing in the Indo-Pacific. TAKAYANAGI *et al.* (1979) and THOMPSON and SCIARRILLO (1978) have noted a prominent first occurrence of *G. conglomerata* in the Brunhes sediments of the Pacific following a long absence of the species since the Pliocene.

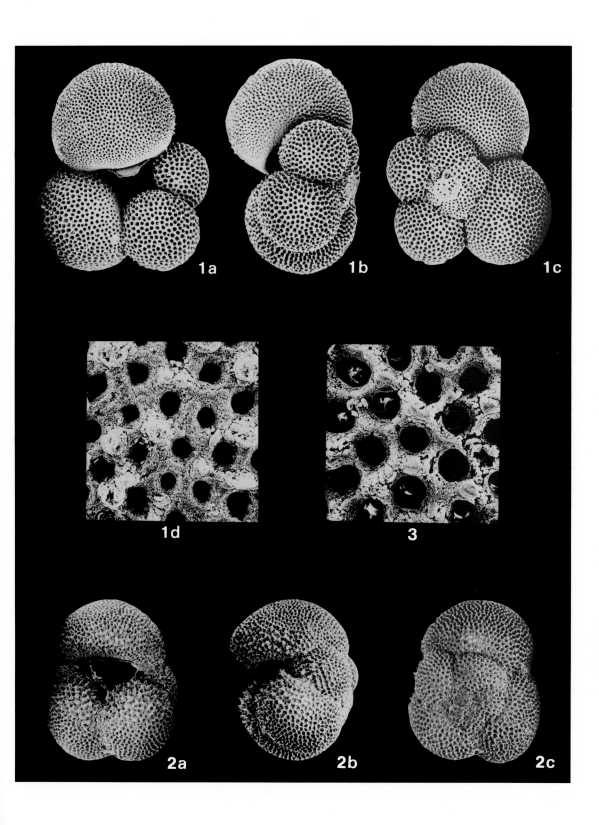

30 *Pulleniatina obliquiloculata* (PARKER and JONES), 1865 —
juvenile forms

Plate 30, Figs. 1a–c, 2a–c, × 133; 1d, 2d, × 1166

Globigerina antilliensis BERMÚDEZ, 1961, p. 1156, pl. 1, figs. 1a–c.
Globigerina atlantis BERMÚDEZ, 1961, p. 1158, pl. 1, figs. 3a–c.
Globigerina santamariaensis McCULLOCH, 1979, p. 416, pl. 172, fig. 7.

Types: Holotype (*antilliensis*)—Atlantis Station 2988 (Lat. 23°15′N, Long. 79°57′W, 380 fm).

Near-Topotype (Figs. 1a–d)—V12-126, trigger core top (Lat. 26°06.9′N, Long. 78°12.4′ W, 902 m).

Holotype (*atlantis*)—Atlantis Station 2988.

Near-Topotype (Figs. 2a–d)—V3-5, core top (Lat. 23°16.7′N, Long. 80°15.5′W, 827 m).

Remarks: These forms are considered to be junior synonyms of *P. obliquiloculata*, representing immature forms differing from *P. obliquiloculata* only in the degree of outer cortex and the extent to which the adult streptospirality has modified the juvenile trochospirality. These forms frequently show a thin apertural rim. A complete gradationally morphologic series can be observed in areas characterized by high numbers of *Pulleniatina,* such as the western Pacific.

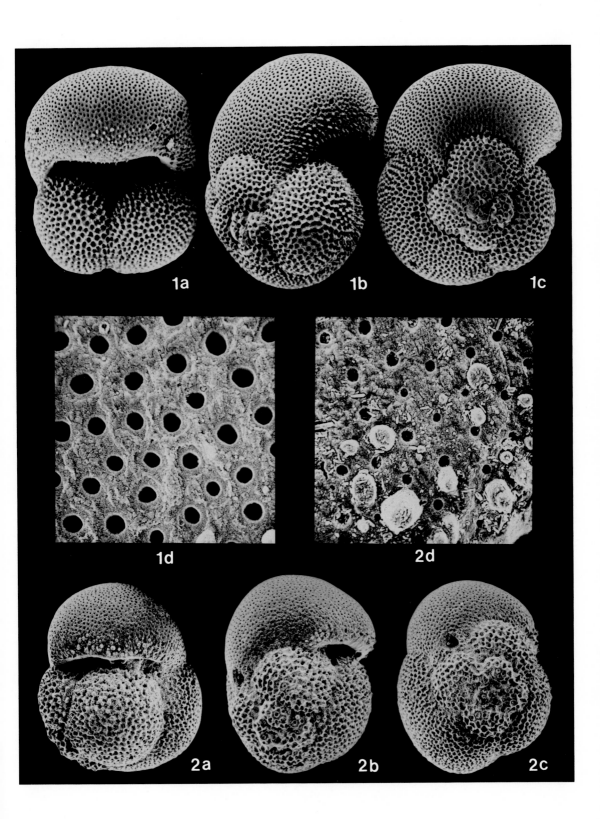

31. 1 *Pulleniatina finalis* BANNER and BLOW, 1967
Plate 31, Figs. 1a–b, 2, × 90; 1c, × 1200

Pulleniatina obliquiloculata (PARKER and JONES) *finalis* BANNER and BLOW, 1967, p. 140, pl. 2, figs. 4–10; pl. 3, fig. 5; p. 4, fig. 10.

Types: Holotype—Challenger Station 344 (Lat. 07°54′30″S; Long. 14°28′20″W, 420 fm).

Plesiotype—V22-170, 9 cm (Lat. 14°38′S, Long. 07°34′W, 4,131 m).

Diagnosis: Test large, initially a low trochospire soon becoming a streptospire, about 4–5 subglobular chambers in the final whorl. Chambers globose, much embracing, increasing moderately in size as added, sutures weak. Aperture a high arch, extraumbilical, extending from the ventral surface near the posterior intercameral suture of the antepenultimate chamber across the periphery of the test to the spiral suture of the antepenultimate chamber, with no apparent lip. Wall calcareous, very smooth and very finely perforated; non-spinose but granular within the umbilicus.

Remarks: BANNER and BLOW (1967, p. 137) recognized this species as having "Umbilical ends of last and opposing chambers meet in a linear suture; umbilical depression broad (never pseudoumbilicate). . . . Dorsally wholly or almost wholly involute; final aperture often almost equally distributed over dorsal and ventral surfaces and is directed dorsally."

Distribution: Early Pleistocene (N. 22) to Recent, equatorial to tropical.

31. 2 *Pulleniatina obliquiloculata* (PARKER and JONES), 1865
Plate 31, Figs. 3a–c, × 90; 3d, × 1200
(see also Pl. 56, Fig. 2).

Pullenia obliquiloculata PARKER and JONES, 1865, p. 183 (*nomen nudum*).

Pullenia sphaeroides (D'ORBIGNY) var. *obliquiloculata* PARKER and JONES, 1865, pp. 365,368; pl. 19, figs. 4a–b. — BANNER and BLOW, 1960a, pp. 25, 40, pl. 7, figs. 4a–c (lectotype).

Pulleniatina obliqueloculata [*sic*] (PARKER and JONES). — CUSHMAN, 1927, p. 90, pl. 19, fig. 5. — BOLLI, LOEBLICH and TAPPAN, 1957, p. 33, pl. 4, figs. 3a–5 (lectotype). — BERMÚDEZ and BOLLI, 1969, pp. 182–183, pl. 16, figs. 1–6.

Globigerina antilliensis BERMÚDEZ, 1961, p. 1156, pl. 1, figs. 1a–c.

Globigerina atlantis BERMÚDEZ, 1961, p. 1158, pl. 1, figs. 3a–c.

Pulleniatina obliquiloculata (PARKER and JONES). — PARKER, 1962, p. 234, pl. 4, figs. 13–16.

Pulleniatina obliquiloculata var. *trochospira* HARTONO, 1964, p. 10, text-figs. a–c.

Pulleniatina obliquiloculata obliquiloculata (PARKER and JONES). — BANNER and BLOW, 1967, pp. 137–139. pl. 3, figs. 4a–c; pl. 4, fig. 9. — BRÖNNIMANN and RESIG, 1971, pp. 1318–1321, pl. 16, figs. 1–11; pl. 17, figs. 1–4, ?5, ?6; pl. 18, figs. 1–7; pl. 19, fig. 6.

Pulleniatina obliquiloculata antilliensis (BERMÚDEZ). — BERMÚDEZ and BOLLI, 1969, p. 183, pl. 16, figs. 7–9.

Pulleniatina okinawaensis NATORI, 1976, pp. 227–228, pl. 5, figs. 5a–c, 6a–c.

Types: Holotype—not given.

Lectotype—Abrolhos Bank (Lat. 22°54′S, Long. 40°37′W, 200 fm), specimen not designated.

Plesiotype—V14-16, trigger core top (Lat. 24°20′S, Long. 41°43′W, 1,928 m).

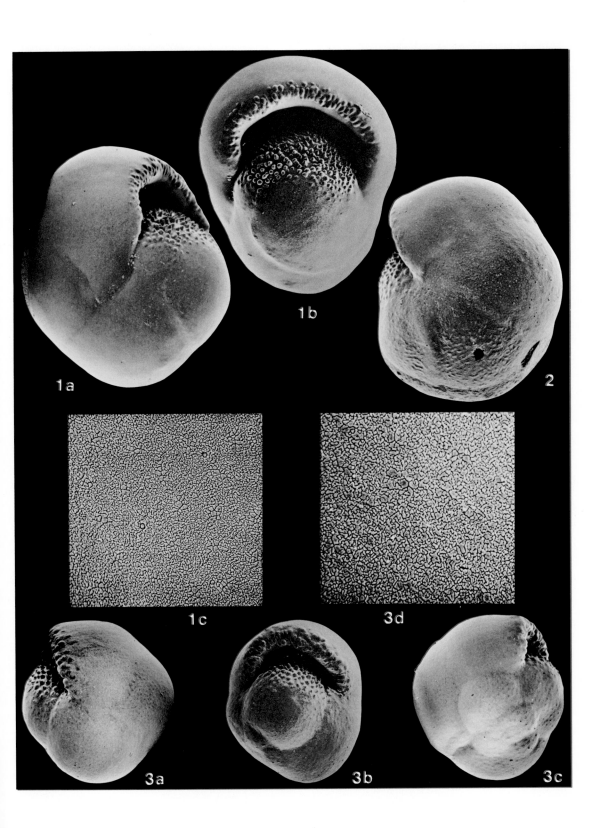

Diagnosis: Test size variable, initially a low trochospire soon becoming a streptospire, 4–4 $1/2$ globular chambers in the final whorl. Chambers spherical becoming radially flattened, much embracing, sutures weak. Aperture a very low, ventrally interiomarginal arch, extending from the periphery of the first chamber of the final whorl to the anterior intercameral suture of the second chamber of the final whorl, with no apparent lip. Wall calcareous, in juveniles coarsely perforate and large circular pores in funnel-like pore depressions, in adults very smooth and nearly perforated; non-spinose but granular in the apertural area.

Remarks: BANNER and BLOW (1967, p. 137) recognized this species as being "dorsally evolute or partially involute; final aperture mainly ventral, usually extends more or less onto dorsal surface, directed approximately normally to original axis of coiling." BANNER and BLOW (1967, p. 138) also called attention to the quite different morphology of immature specimens of *P. obliquiloculata* designated by BERMÚDEZ (1961) as *G. antilliensis* (see Plate 21). We also include *G. atlantis* BERMÚDEZ and *G. santamariaensis* McCULLOCH as representing juvenile *Pulleniatina*. See also BRÖNNIMANN and RESIG (1971) for their very well-illustrated discussion of ontogenetic developments as well as BURT and SCOTT (1975).

Distribution: Early Pliocene (N. 19) to Recent, equatorial and tropical.

32. 1   *Pulleniatina praecursor* BANNER and BLOW, 1967
Plate 32, Figs. 1a, 2a–b, × 90; 1b, × 1200

Pulleniatina obliquiloculata (PARKER and JONES) *praecursor* BANNER and BLOW, 1967, p. 139, pl. 3, fig. 3.

Types: Holotype—STAINFORTH Sample 19305, Borbón Formation, north coast of Ecuador between Rio Chevele and Estero Cienago.

Plesiotype—V16-205, 960 cm (Lat. 15°24′N, Long. 43°24′W, 4,043m).

Diagnosis: Test medium to large, initially trochospiral soon becoming streptospiral, about 5 chambers in the final whorl. Chambers subglobular, increasing moderately in size as added, greatly embracing, sutures weak. Aperture a very low interiomarginal arch, extending from the periphery of the first chamber in the final whorl almost to the anterior intercameral suture of the second chamber of the final whorl, with no apparent lip. Wall calcareous, smooth and very finely perforated, non-spinose but granular around the aperture.

Remarks: BANNER and BLOW's (1967, p. 137) key to the species of the genus *Pulleniatina* separated *P. praecursor* as having "umbilical ends of late chambers meet at a point in a very narrow umbilical depression (may be pseudoumbilicate); final aperture does not extend onto dorsal surface of penultimate whorl."

Distribution: Early Pliocene (N. 19) to Early Pleistocene (N. 22), equatorial.

32. 2   *Pulleniatina primalis* BANNER and BLOW, 1967
Plate 32, Figs. 3a–c, × 90; 3d, × 1200

Pulleniatina semiinvoluta PARKER (not GERMERAAD, 1946), 1965b, pp. 151,152, text-figs. 5a–6c.
Pulleniatina primalis BANNER and BLOW, 1967, p. 142, pl. 1, figs. 3–8; pl. 3, fig. 2.

Types: Holotype—Sample WHB 181B, Bowden Formation, Buff Bay, Jamaica.

Plesiotype—V16-205, 1,078 cm (Lat. 15°24′N, Long. 43°24′W, 4,043 m).

Diagnosis: Test medium to large, initially a low trochospire soon becoming a streptospire, $4^1/_2$–5 chambers in the final whorl, about 14 chambers arranged in about 3 whorls. Chambers subglobular, increasing moderately in size as added, much embracing, sutures weak. Aperture a low, interiomarginal arch, extending from the periphery of the first chamber of the last whorl to the anterior intercameral suture of the antepenultimate chamber of the last whorl, with no apparent lip. Wall calcareous, very smooth and very finely perforated; non-spinose but granular in the umbilicus.

Remarks: BANNER and BLOW (1967, p. 137) recognized this species as having its "aperture confined to the inner, more umbilical part of the final basal suture, not reaching the periphery of the preceding whorl; streptospirality only ventrally directed; ventral intercameral sutures meet at a point in the umbilical depression . . . periphery broadly rounded, not subacutely angled or pseudocarinate."

Distribution: Late Miocene (N. 17) to Early Pleistocene (N. 22), equatorial.

Plate 32 *Pulleniatina praecursor* BANNER and BLOW, 1967
Pulleniatina primalis BANNER and BLOW, 1967

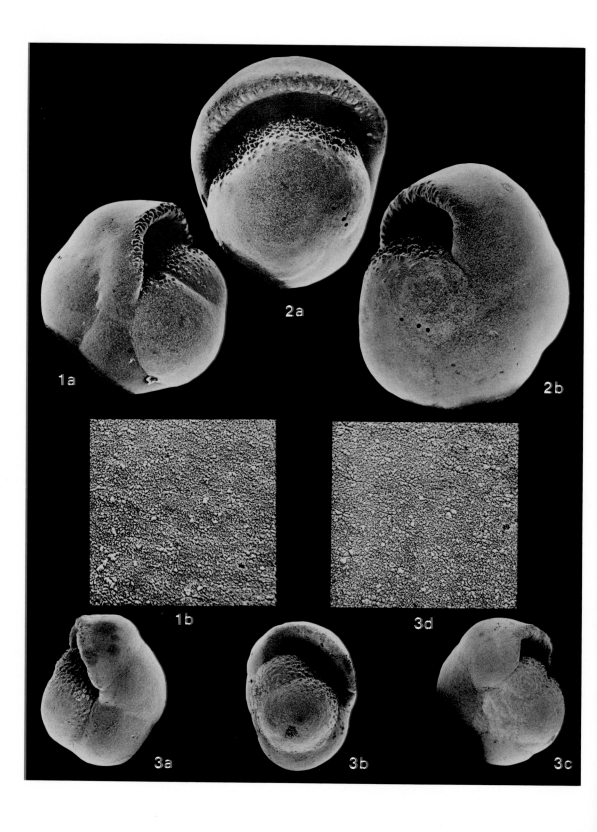

33. 1 *Neogloboquadrina pseudopachyderma* (CITA, PREMOLI-SILVA and ROSSI), 1965

Plate 33, Figs. 1a–c, × 250; 1d, × 2083

Globigerina pachyderma (EHRENBERG). — TAKAYANAGI and SAITO, 1962, p. 89, pl. 26, figs. 4a–c. — PEZZANI, 1963, pp. 585–586, pl. 30, figs. 6a–b; pl. 31, figs. 12a–b. — PARKER, 1964, p. 630, pl. 101, figs. 12–14.

Globorotalia pseudopachyderma CITA, PREMOLI-SILVA and ROSSI, 1965, pp. 233–236, pl. 20, figs. 3a–c, 4a–c, 6; pl. 31, figs. 6a–c; p. 235, text-figs. 5c–d.

Globorotalia (Turborotalia) pseudopachyderma CITA, PREMOLI-SILVA and ROSSI. — OLSSON, 1974, pp. 47–60, pl. 2, figs. A-J; pl. 3, figs. H–Q.

Neogloboquadrina pseudopachyderma (CITA, PREMOLI-SILVA and ROSSI). — POAG and VALENTINE, 1976, p. 203, pl. 13, fig. 4.

Types: Holotype (*G. pseudopachyderma*)—Level 12, type Tortonian section, Rio Mazzapiedi Series near Castelliana, Tortona-Alessandria region, northern Italy (Miocene).

Plesiotype——RC11-191, 1,030 cm (Lat. 44°31′N, Long. 139°56′W, 4,387 m).

Diagnosis: Test small to medium in size, 4 globular chambers in the final whorl, about 13 chambers in all arranged in a low trochospire of about 3 whorls, chambers spherical to subspherical, closely packed, size increases slowly as added giving only slight lobulation, sutures depressed, nearly radial. Aperture umbilical-extraumbilical, a low interiomarginal arch, partially obscured by a fairly thick rim-like lip at the base of the final chamber. Wall calcareous, perforated, non-spinose but coarsely pustulate.

Remarks: Since the *N. pachyderma* group represents probably the only commonly occurring planktonic foraminiferal taxa in cold temperate to polar latitudes, considerable study has been made of them. A major conclusion is that Pliocene and early Pleistocene sediments contain the form *N. pseudopachyderma* which is only partially conspecific with Quaternary *N. pachyderma*. The position of the aperture is more extraumbilical in *N. pseudopachyderma* and the chambers are more inflated.

Distribution: Late Miocene (N. 16) to early Pleistocene (N. 22), cold temperate waters.

33. 2 *Neogloboquadrina cryophila* (HERMAN), 1980

Plate 33, Figs. 2a–b, 3, × 250; 2c, × 2083

Globigerina sp. HERMAN, 1969, pl. 1, figs. 13–16.

Globigerina occlusa HERMAN, 1974, pp. 300–301, pl. 3, figs. 13–16; pl. 10, figs. 3–4 (invalid name because of homonymy with *G. praebulloides occlusa* BLOW and BANNER, 1962).

Globigerina cryophila HERMAN, 1980, p. 631 (new name).

Types: Holotype (*G. cryophila* HERMAN)—From Drift Station Alpha 6, 58.5–59.5 cm (Lat. 85°15′N, Long. 167°54′W, 1,842 m).

Topotype—From Drift Station Alpha 6, 59 cm.

Diagnosis: Test small, 4 subglobular chambers in the final whorl, about 10 chambers in all arranged in a low trochospire of about $2^1/_2$ whorls, chamber size increases slowly as added, chambers much embracing giving only weak lobulation to the equatorial periphery, sutures well incised, radial. Aperture umbilical, a subquadrate opening beneath a stout lip on the final chamber. Wall calcareous, perforated, coarsely cancellate, non-spinose.

Remarks: It is difficult to characterize this taxon from the great morphologic variability illustrated by HERMAN (1969, 1974). Specimens collected from the holotype level are considered to be conspecific with *N. pachyderma*, differing only in the tangentially inflated chambers and more umbilical aperture.

Distribution: Noted only from the Quaternary of the Arctic Ocean.

Plate 33 *Neogloboquadrina pseudopachyderma* (CITA, PREMOLI-SILVA and ROSSI), 1965
Neogloboquadrina cryophila (HERMAN), 1980

105

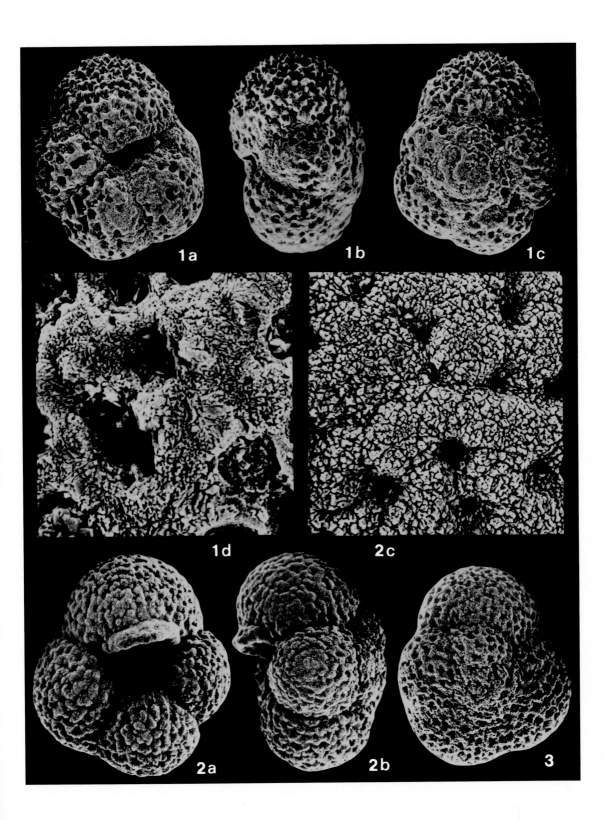

34. 1   *Neogloboquadrina pachyderma* (EHRENBERG), 1861
Plate 34, Figs. 1a–c, × 133; 1d, × 1333
(see also Pl. 56, Fig. 6)

Aristerospira pachyderma EHRENBERG, 1861, pp. 276,277,303 — EHRENBERG, 1872, pl. 1, fig. 4.
Globigerina bulloides D'ORBIGNY, "Arctic variety," BRADY, 1878, p. 435, pl. 21, figs. 10a–c.
Globigerina bulloides D'ORBIGNY var. *borealis* BRADY, 1881, p. 412. — BANNER and BLOW, 1960, p. 4, pl.
    3, figs. 4a–c (lectotype).
Globigerina pachyderma (EHRENBERG). — BRADY, 1884, p. 600, pl. 114, figs. 19–20. — BÉ, 1960,
    p. 66, text-fig. 1. — KENNETT, 1968b, pp. 305–318, pl. 1, figs. 1–32. — HERB, 1968, pp. 473, 475,
    fig. 3; pl. 3, figs. 1–2. — BOLTOVSKOY, 1969, p. 248, pl. 1, figs. 11a–c. — KENNETT, 1970,
    pp. 47–49, pl. 8, figs. 1–12; pl. 9, figs. 1–12. — PARKER, 1962, pp. 224–225, pl. 1, figs. 26–35; pl. 2,
    figs. 1–6. — OLSSON, 1975, pp. 244–257, pls. 1–6.
Globigerina borealis BRADY. — BLOW, 1969, p. 312. — AKERS, 1972, pp. 45–46 — ASANO, 1957,
    pp. 2, 12–14, 25, 26; pl. 1, figs. 16–18.
Neogloboquadrina pachyderma pachyderma (EHRENBERG). — RÖGL and BOLLI, 1973, p. 571, pl. 11,
    figs. 2–6; pl. 16, fig. 12.
Turborotalia (Turborotalia) pachyderma (EHRENBEG). — FLEISHER 1974a, p. 1036, pl. 19, fig. 13.
Globorotalia (Turborotalia) pachyderma (EHRENBERG). — BANDY, 1972b, pp. 294–318, pls. 1–8.
Globorotalia pachyderma pachyderma (EHRENBERG). — IACCARINO and SALVATORINI, 1979, p. 264,
    pl. 7, figs. 10, 11.
Globorotalia pachyderma borealis (BRADY). — IACCARINO and SALVATORINI, 1979, p. 264, pl. 7, figs.
    15–18.

Types: Holotype—Lat. 62°40′N, Long. 29°W (see PARKER, 1962, p. 224).

Holotype (*borealis*)— Knight Errant Station 8 (Lat. 60°03′N, Long. 05°51′W,
5,405 m).

Lectotype (*borealis*)—North-Polar Expedition, 1875–1876, northernmost
sounding, 83°19′N, 72 fm (Fig. 10, BRADY, 1878).

Plesiotype——V23-35, trigger core top (Lat. 62°00.2′N, Long. 28°41.4′W,
1,624 m).

Diagnosis: Test small, 4–5 globular chambers in the final whorl, about 14 chambers
arranged in 3 whorls. Chambers subspherical to ovate, closely embracing, size increasing fairly rapidly as added, final chamber frequently forming a bulla-like cover
over much of the umbilicus. Aperture umbilical, interiomarginal, a low arch with a
thick and wide apertural rim protruding slightly above the level of the final chamber.
Wall calcareous, thick, granular, pores frequently multiple in a single pore depression,
non-spinose.

Remarks: This species can be distinguished from *N. incompta* by its more numerous
and more embracing chambers, and from small specimens of *N. eggeri* by the position
of the apertural rim which is well out of the umbilicus. Our examination of the lectotype
of *G. borealis* BRADY suggested that the apertural lip was prominent than figured by
BANNER and BLOW, and it seemed possible that the specimen was not *N. pachyderma*.
After examining variation in surface ultramicrostructures of *N. pachyderma* from the
South Pacific, SRINIVASAN and KENNETT (1975) believe that this species builds a relatively unthickened, pitted microcrystalline surface (termed "reticulate forms") in warm
water masses and a more thickened crystalline surface covered by distinct euhedral
calcite rhombs (termed "crystalline forms") in cool water masses.

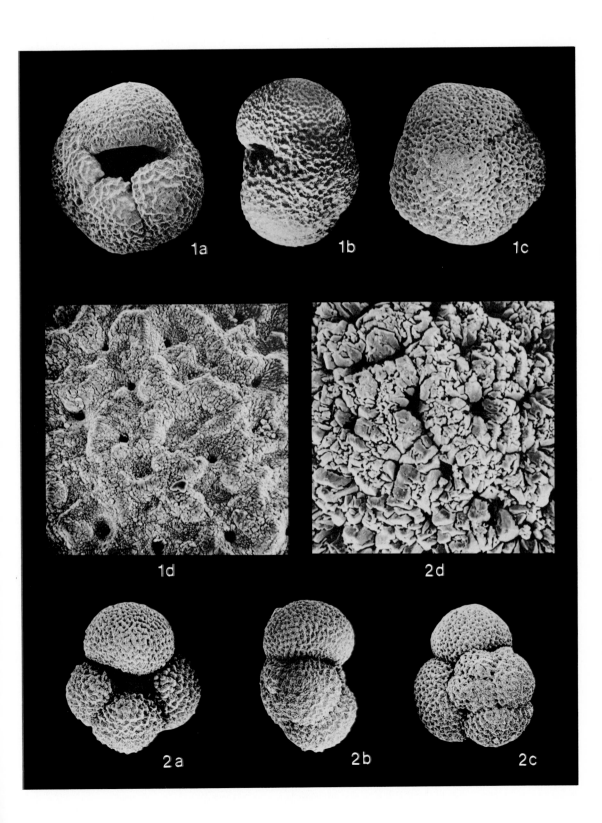

Distribution: At high latitudes in both hemispheres, this species is monospecific. Kennett (1968b) noted that 4-chambered forms with a very quadrate outline and dominantly left coiling were the most polar varieties, giving way to left-coiling $4^1/_2$-chambered forms, and then to $4^1/_2$-chambered right-coiling forms with decreasing latitude. Ericson (1959) and Bandy (1960) noted the change from dominantly left-coiling to dominantly right-coiling occurred at the 7.2°C April isotherm. *N. pachyderma* has been observed in equatorial Pacific sediments (Thompson, 1976).

34.2 *Neogloboquadrina incompta* (Cifelli), 1961
Plate 34, Figs. 2a–c, × 133; 2d, × 1333

Globigerina incompta Cifelli, 1961, p. 83, pl. 4, figs. 1–7. — Cifelli and Smith, 1970, pp. 26–28, fig. 3; text-fig. 14. — Cifelli, 1973, pp. 157–166, pl. 2, figs. 5–12; pl. 3, figs. 1,6,10; pl. 4, figs. 1,3.
Globigerina pachyderma incompta Cifelli. — Cifelli, 1965, p. 11, pl. 1, figs. 4–6.
Globigerina pachyderma (Ehrenberg) forma *superficiaria* Boltovskoy, 1969, p. 248, pl. 1, figs. 12a–b; pl. 2, fig. 1.
Neogloboquadrina pachyderma incompta (Cifelli). — Rögl and Bolli, 1973, p. 571, pl. 10?

Types: Holotype—R/V Crawford Station 3 (Lat. 38°39′N, Long. 69°33′W, plankton haul).

Near-Topotype—V20-253, trigger core top (Lat. 38°17′N, Long. 68°32′W, 4,889 m).

Diagnosis: Test small in size, about 4–$4^1/_2$ globular chambers in the final whorl, about 9–14 chambers in all arranged in about 2–3 trochospiral whorls, equatorial periphery subquadrilateral, chambers subspherical, partly embracing, size increasing moderately as added, except for the last 2–3 chambers which are added rapidly, sutures distinct, final chamber slightly overhanging the umbilicus toward the ventral side. Aperture umbilical, interiomarginal but extending nearly to the periphery, a low arch with a thin projecting rim. Wall calcareous, thin, finely porous, granular, spinose(?).

Remarks: The specimen illustrated here conforms to what Cifelli (1973) describes as the "thick [walled] bottom sediment form" of *N. incompta*, which also has a thin-walled counterpart in North Atlantic sediments. It is differentiated from *N. pachyderma* by its more coarsely pitted and more pustulate surface, and, in samples from higher latitudes, is more lobulate when found in association with *N. pachyderma*. Spines are reported on plankton tow specimens (Cifelli, 1961), but such forms were not illustrated, and the material studied here did not reveal spines or spine bases.

Distribution: Cifelli (1961) reports spinose representatives from surface plankton tows in the North Atlantic. Shinbo and Maiya (1970) used it as a biostratigraphic zonal fossil in latest Pliocene and Pleistocene sediments from northern Japan.

35 *Neogloboquadrina polusi* (ANDROSOVA), 1962
Plate 35, Figs. 1a–c, 2a–c, × 135; 1d, 2d, × 1200

Globigerina polusi ANDROSOVA, 1962, pp. 113–114, text-figs. 15a–c.
Globigerina sp. HERMAN, 1969, p. 269, pl. 1, figs. 8–12.
Globigerina paraobesa HERMAN, 1974, p. 305, pl. 3, figs. 8–12; pl. 11, figs. 1–4; pl. 12, figs. 1–4.

Types: Syntype (*N. polusi*)— Severnyi Polyus I Drift Station (1937–1938);
Sta. 4 (Lat. 88°47′N, Long. 10°01′W, 4,290 m) core top;
Sta. 15 (Lat. 86°09′N, Long. 00°58′E, 3,767 m) 15–20 cm;
Sta. 16 (Lat. 85°40′N, Long. 01°02′E, 4,050 m) 5–7 m.
Plesiotypes—T3-67-12, 0–3 cm (Lat. 80°21.9′N, Long. 173°33′W, 2,867 m).
Holotype (*G. paraobesa*)—Drift Station Alpha 3, 10 cm (Lat. 84°12′N, Long. 168°33′W, 2,409 m).

Diagnosis: Test medium in size, a low to moderate-height trochospire, 5–$5^1/_2$ subspherical chambers in the final whorl, about 12–15 rapidly enlarging chambers arranged in $2^1/_2$–3 whorls, sutures deep, radial. Aperture umbilical-extraumbilical, a high and wide arch occasionally showing a lip on the final chamber and secondary openings on the dorsal side. Wall calcareous, perforated, non-spinose, coarsely cancellate.

Remarks: Although characterized by large, umbilically-extended chambers, we believe that this form is related to *N. pachyderma* and probably conspecific with it. An unpublished study of plankton tow samples in the Arctic by SAITO shows this form to occur seasonally in low-salinity waters near melting ice. It should therefore be regarded as an ecologic variant rather than as a separate species.

Distribution: Noted only in Quaternary sediments of the Arctic Ocean.

Plate 35 *Neogloboquadrina polusi* (ANDROSOVA), 1962

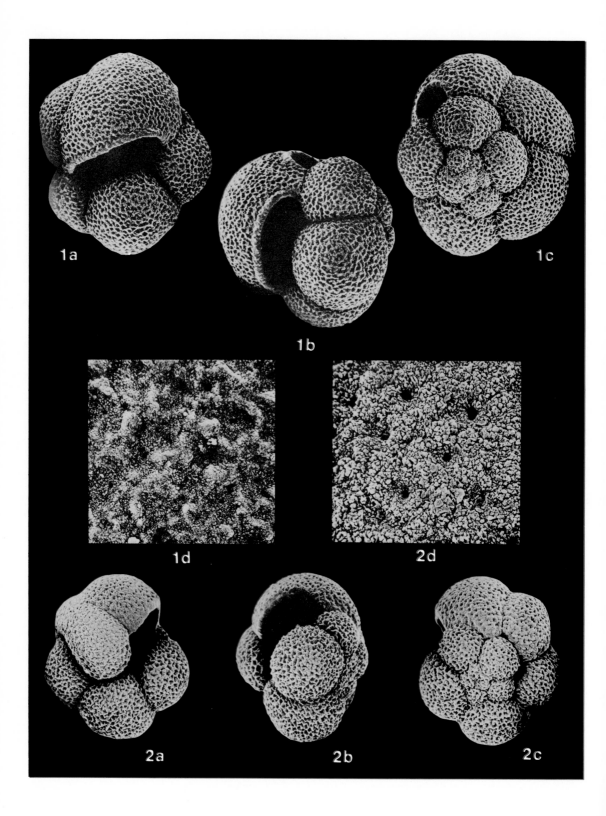

36. 1 *Neogloboquadrina dutertrei* (D'ORBIGNY), 1839
Plate 36, Figs. 1a–c, × 66; 2, × 666

Globigerina rotundata D'ORBIGNY, 1826, p. 277, list. no. 6 (*nomen nudum*). — FORNASINI, 1899, p. 208, text-fig. 3 (after D'ORBIGNY) (lectotype). — BANNER and BLOW, 1960a, p. 19, pl. 2, fig. 2.

Globigerina dutertrei D'ORBIGNY, 1839a, p. 84, pl. 4, figs. 19–21 (lectotype). — BANNER and BLOW, 1960a, p. 11, pl. 2, fig. 1. — CIFELLI, 1965, p. 12, pl. 2, figs. 1–2. — CIFELLI and SMITH, 1970, p. 21, p. 2, figs. 1–2, text-fig. 13.

Globoquadrina dutertrei (D'ORBIGNY). — PARKER, 1962 (part), p. 242, pl. 7, figs. 1–5 (not pl. 7, figs. 6–13; pl. 8, figs. 1–4).

Neogloboquadrina dutertrei dutertrei (D'ORBIGNY). — RÖGL and BOLLI, 1973, pl. 9, figs. 1–3, 7–10; pl. 17, figs. 1–6.

Globigerina galapagosensis McCULLOCH, 1979, p. 413, pl. 171, fig. 8.

Globigerina hancocki McCULLOCH, 1979, p. 413, pl. 172, fig. 6.

Globigerina hybrida McCULLOCH, 1979, p. 414, pl. 175, figs. 4–7, 15.

Types: Syntypes (*dutertrei*)—Recent marine sands of Cuba, Martinique and Guadaloupe.

Syntypes (*rotundata*)—Saint Helena Island.

Lectotype—Remaining whole specimen from Cuban specimens.

Plesiotype—V3-3, trigger core top (Lat. 18°51′N, Long. 67°07′W, 2,661 m).

Diagnosis: Test medium to large, a medium-height trochospire, 4–5 globular chambers in the final whorl, about 12 chambers in all arranged in about 3 whorls. Chambers subspherical to slightly flattened radially, increasing slowly in size as added, sutures distinct, depressed. Aperture umbilical, interiomarginal, a wide and deep opening directly into the umbilicus, with no apparent lip, but may have an umbilical tooth. Wall calcareous, moderately perforated, surface coarsely pitted but non-spinose.

Remarks: Although BANNER and BLOW (1960, p. 11) referred this species to *Globigerina*, SEM study shows it to be nonspinose. PARKER (1962, p. 24) switched it to *Globoquadrina* because of its pitted surface texture, lack of spines and occasional umbilical tooth. She also felt, along with ZOBEL(1968), that the various forms, referred to here as *N. eggeri* and *N. blowi*, were phenotypes of *N. dutertrei*. It is the opinion of the authors here that these three, as well as additional forms figured in later plates, can be separated morphologically and more importantly, phylogenetically. *N. dutertrei* is more high spired than either *N. eggeri* or *N. blowi*. The presence or absence of an umbilical tooth is not a deciding criterion for the recognition of the species, and may be environmentally or preservationally controlled.

Distribution: Middle Pliocene (N. 20) to Recent—Caribbean? only—not observed in the Pacific (THOMPSON, 1976).

36. 2 *Neogloboquadrina eggeri* (RHUMBLER), 1901
Plate 36, Figs. 3a–c, × 66; 4, × 666

Globigerina dubia BRADY (not EGGER, 1857) — BRADY 1879, p. 285. — BRADY, 1884, p. 595, pl. 79, figs. 17a–c. — FLINT, 1899, p. 332. — CUSHMAN, 1914, pp. 6–7, pl. 4, figs. 1–3.

Globigerina eggeri RHUMBLER, 1901, pp. 12–20 text-fig. 20 (after Brady). — BANNER and BLOW, 1960a, pp. 11–12, pl. 2, figs. 4a–c. (lectotype).

Globigerina pyriporosa RHUMBLER, 1911, pp. 134,137, pl. 31, figs. 1–3, 4a–b.

Globoquadrina dutertrei (D'ORBIGNY). — PARKER, 1962. p. 242, pl. 7, figs. 6–12; pl. 8, figs. 1–4.

Neogloboquadrina dutertrei (D'ORBIGNY) subsp. *dutertrei* (D'ORBIGNY). — BANDY, FRERICHS and
VINCENT, 1967, pp. 152–157, pl. 14, figs. 2?, 3–12.
Neogloboquadrina dutertrei dutertrei (D'ORBIGNY). — RÖGL and BOLLI, 1973, p. 570, pl. 9, figs. 3,6;
pl. 10, figs. 1–10; pl. 17, fig. 12.
Turborotalia (Neogloboquadrina) dutertrei (D'ORBIGNY). — FLEISHER, 1974a, p. 1035, pl. 18, fig. 5.
Globigerina eggeriformis MCCULLOCH, 1979, pp. 412–413, pl. 171, fig. 9.

Types: Syntype (*dubia*)—Challenger Station 300 (Lat. 33°42'S, Long. 78°18'W,
2,515 m).

Syntype (*eggeri*)—North and South Atlantic, North and South Pacific, 56°N
–46°S.

Lectotype—Original materials of BRADY (1884).

Plesiotype (*eggeri*)—V15-55, trigger core top (Lat. 34°36.5'S, Long. 76°07'W,
3,941 m).

Diagnosis: Test medium to large in size, a low trochospire, 5–7 globular chambers
in the final whorl, about 15–18 chambers in all arranged in $2^1/_2$–3 whorls. Chambers
spherical to slightly flattened radially, increasing slowly in size as added, closely
embracing, sutures distinct, deep. Aperture umbilical, interiomarginal, a wide and
deep opening, with or without umbilical teeth. Wall calcareous, perforated, coarsely
pitted, non-spinose.

Remarks: The morphology of this species is quite variable, undoubtedly due to its
wide range of habitats. Equatorial and tropical representatives tend to be larger with
fewer chambers than temperate forms. The presence or absence of umbilical teeth is
probably controlled more by environment or preservation than it is a deciding criterion
for recognition of the species. *N. eggeri* has a trochospire which is higher than *N. blowi*
but lower than *N. dutertrei*, and a more umbilical aperture.

Distribution: Early Pleistocene to Recent, equatorial to cool temperate.

Plate 36 *Neogloboquadrina dutertrei* (D'ORBIGNY), 1839
Neogloboquadrina eggeri (RHUMBLER), 1901

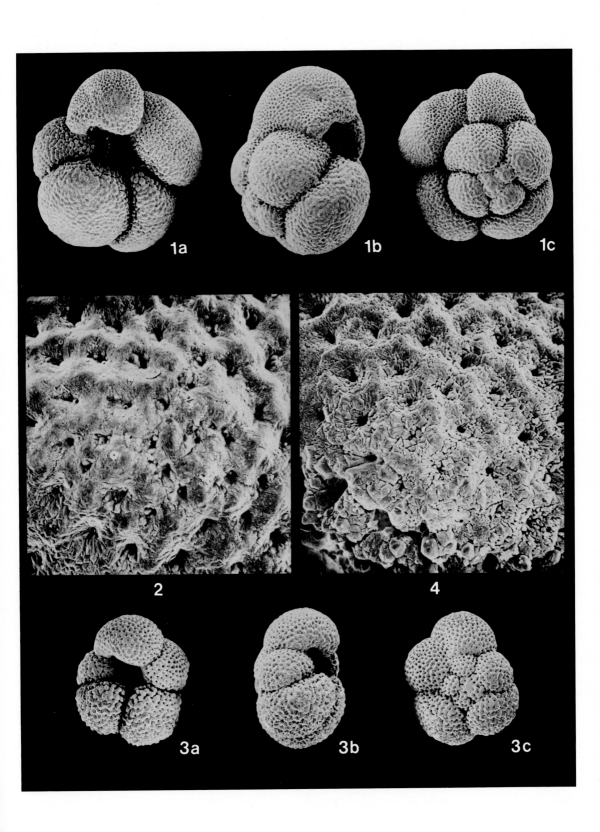

37. 1   *Neogloboquadrina blowi* RÖGL and BOLLI, 1973
Plate 37, Figs. 1a–c, × 120; 1d, × 600

Globigerina cretacea BRADY, 1884 (not D'ORBIGNY), p. 596, pl. 82, fig. 10.
Globigerina cretacea LOMNICKI, 1900 (not D'ORBIGNY), pl. 1, figs. 2a–c.
Globigerina subcretacea LOMNICKI, 1901, p. 17.
cf. *Globigerina subcretacea* CHAPMAN, 1902, p. 410, pl. 36, figs. 16a–b.
Globigerinella subcretacea (CHAPMAN).   —   THALMANN, 1932, p. 307.
Globigerinella subcretacea (LOMNICKI).   —   THALMANN, 1933, p. 253.
Globorotalia subcretacea (LOMNICKI).   —   TAKAYANAGI and SAITO, 1962, p. 81, pl. 24, figs. 8a–c.
Neogloboquadrina dutertrei (D'ORBIGNY) subsp. *subcretacea* (LOMNICKI).   —   BANDY, FRERICHS and
   VINCENT, 1967, pp. 152–157, pl. 14, fig. 2.
Globigerina dutertrei D'ORBIGNY forma C, ZOBEL, 1968, p. 111, abb. 3.
Globorotalia (Turborotalia) subcretacea (LOMNICKI).   —   BLOW, 1969, p. 392, pl. 4, figs. 18–20 (lectotype of
   Globigerina cretacea BRADY).
Neogloboquadrina dutertrei blowi RÖGL and BOLLI, 1973, p. 570, pl. 9, figs. 15–18, 19–20? 21–22; pl. 17, fig.
   12 (new name for *Globigerina subcretacea* CHAPMAN, 1902).

Types: Syntypes (LOMNICKI)——several Polish Miocene localities, and Recent after
   BRADY, 1884.

Syntypes (CHAPMAN)——200 fm off Tutanga, near Funafti.

Lectotype (LOMNICKI after BRADY, 1884)—Challenger Station 191A (Lat.
   05° 26′S, Long. 133° 19′E, 580 fm).

Plesiotype—V28-333, trigger core top (Lat 03° 39′S, Long. 132° 18′E, 2,125 m).

Diagnosis: Test medium to large in size, very low trochospire, 4$^1/_2$–5 globular chambers
in the final whorl, about 10–12 chambers in all arranged in 2$^1/_2$ whorls. Chambers
spherical increasing rapidly in size as added, chambers slightly embracing, sutures
distinct and deep umbilical tooth. Wall calcareous, finely perforated, moderately pitted.

Remarks: The taxonomic history of this species has been very confused, due to con-
tinual misidentification by later workers of older material. This finally caused RÖGL and
BOLLI (1973, p. 570), to propose a new name for *Globigerina subcretacea* CHAPMAN, 1902,
an invalid homonym for *Globigerina subcretacea* LOMNICKI, 1900, which, in turn, was
an incorrect synonym for the form identified incorrectly by BRADY, 1884 as *Globigerina
cretacea* D'ORBIGNY.

The placement of *N. blowi* into *Neogloboquadrina* is made on the basis of its pitted
surface, similar to, but less coarse than, those of *N. dutertrei* and *N. eggeri*. It could,
however, be placed into *Globorotalia*, as was done by BLOW (1969), because of its
umbilical-extraumbilical aperture. The trochospire, in any case, is very low, often
depressed below the final whorl of, typically, 4$^1/_2$ greatly enlarged chambers. This, plus
the quite lobulate equatorial profile, distinguishes it from *N. blowi, N. eggeri* and *N.
humerosa*.

37. 2   *Neogloboquadrina humerosa* (TAKAYANAGI and SAITO), 1962
Plate 37, Figs. 2a–c, × 120; 2d, × 600

Globorotalia humerosa TAKAYANAGI and SAITO, 1962, p. 78, pl. 28, fig. 1–2.
Globoquadrina humerosa (TAKAYANAGI and SAITO).   —   PARKER, 1967, pp. 169–170, pl. 24, figs. 10, 11;
   pl. 25, figs. 1–6.

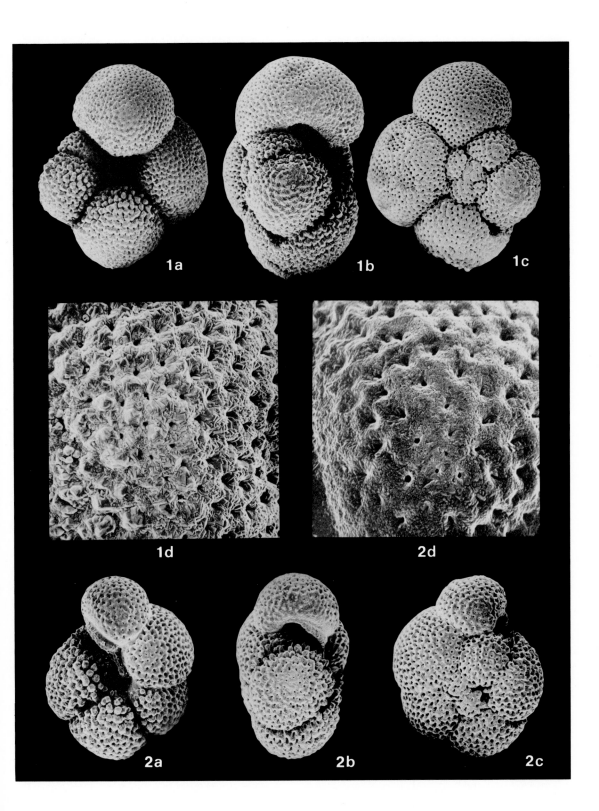

Globorotalia (Turborotalia) acostaensis humerosa (TAKAYANAGI and SAITO). — BLOW, 1969, pp. 345–346, pl. 33, figs. 4, 5, 7–9; pl. 34, figs. 1–3.

Globorotalia dutertrei humerosa (TAKAYANAGI and SAITO). — BOLLI, 1970, p. 580, pl. 2, figs. 4–6.

Globorotalia (Turborotalia) dutertrei (D'ORBIGNY). — JENKINS, 1971, p. 114.

Turborotalia humerosa (TAKAYANAGI and SAITO). — POAG, 1972, p. 512, pl. 2, figs. 9–10.

Turborotalia (Turborotalia) humerosa humerosa (TAKAYANAGI and SAITO). — FLEISHER, 1974a, p. 1035, pl. 19, fig. 4.

Types: Holotype—Sample A-16, Nobori Formation, Nobori, Muroto City, Kochi Prefecture Japan (Lat. 33°22′09″N, Long. 134°03′33″E).

Topotype—Original materials, Sample A-9 Nobori Formation.

Diagnosis: Test medium to large in size, 6–7 globular chambers in the final whorl, 10–14 chambers in all arranged in about 2 whorls. Chambers spherical, slowly enlarging in size as added, typically one or more chambers in the final whorl are of a size out of the increasing progression, so that the whorl becomes distorted. The final chamber is displaced towards the umbilical side. Sutures distinct on the umbilical side, weak on spiral side. Aperture umbilical-extraumbilical, a long narrow circle, narrowed further by a typically occurring thick apertural rim which runs from chamber to chamber in the final whorl. Wall calcareous, perforated, moderately pitted.

Remarks: TAKAYANAGI and SAITO (1962) differentiated this form from *N. eggeri* by its umbilical-extraumbilical aperture and apertural lips, and from *N. subcretacea* by its less lobulated periphery and irregularly developed final chambers. It is here included in *Neogloboquadrina* because of its surface pitting.

Distribution: Middle Pliocene (N. 20) to Late Pleistocene (N.21), tropical. THOMPSON and SCIARRILLO (1978) placed the extinction level of this species at 1.0 million years B.P.

38. 1 *Neogloboquadrina asanoi* (MAIYA, SAITO and SATO), 1976
Plate 38, Figs. 3a–c, × 90; 4, × 600

Globigerina conglomerata SCHWAGER. — TAKAYANAGI and SAITO, 1962 (not SCHWAGER, 1866), p. 84, pl. 24, figs. 10a–c.
Globoquadrina asanoi MAIYA, SAITO and SATO, 1976, p. 409, pl. 3, figs. 1a–c, 2a–c, 3.
Neogloboquadrina asanoi MAIYA, SAITO and SATO. — THOMPSON, 1980, pl. 3, figs. 10–12.

Types: Holotype and Paratype—Sample N. N. 1, upper part of Funakawa Formation (Nomura Mudstone Member), seaside cliff on northern coast of Oga Peninsula, Akita Prefecture, Northwest Japan.

Diagnosis: Test medium to large in size, a low trochospire with 4 globular chambers in the final whorl, about 10–11 chambers in all arranged in about $2^{1}/_{2}$ whorls. Chambers subspherical to ovate, fully inflated, partially embracing, size increasing gradually as added, sutures well depressed. Aperture interiomarginal, umbilical, a low arch with no apparent modifications. Wall calcareous, coarsely pitted, non-spinose.

Remarks: This species was separated by its authors from *G. conglomerata* by its "protruding conical arrangement of the early chambers on the spiral side and by its more globular, inflated chambers."

Distribution: Middle part of the Gauss to the base of the Olduvai Magnetic Event, probably in temperate waters.

38. 2 *Neogloboquadrina kagaensis* (MAIYA, SAITO and SATO), 1976
Plate 38, Figs. 1a–c, × 90; 2, × 600

Globoquadrina kagaensis MAIYA, SAITO and SATO, 1976, pp. 409–410, pl. 3, figs. 4a–b, 5, 6a–c.

Types: Holotype—Sample N-YB-65, from lower part of Yabuta Formation, seacliff on eastern coast of Noto Peninsula, 1 km north of Ozaki, Himi City, Toyama Prefecture, central Japan.

Paratype—Core RC10-161, 940 cm (Lat. 33°05′N, Long. 158°00′E, 3,587m)

Diagnosis: Test medium to large in size, 4–$4^{1}/_{2}$ globular chambers in the final whorl, about 14 to 15 chambers in all arranged in about $2^{1}/_{2}$ whorls. Chambers subspherical to ovate, partially embracing, increasing gradually in size as added, sutures well depressed. Aperture interiomarginal, umbilical-extraumbilical, a low arch with a distinct tooth-like flap. Wall calcareous, non-spinose, granular.

Remarks: This species is directly descended from *N. asanoi*, and can be distinguished by its more extraumbilical aperture, more quadrate equatorial periphery and $4^{1}/_{2}$ chambers in the final whorl instead of 4 as in *N. asanoi*.

Distribution: Late Pliocene to Early Pleistocene, temperate waters of the northwest Pacific.

38. 3 *Neogloboquadrina himiensis* (MAIYA, SAITO and SATO), 1976
Plate 38, Figs. 5a–c, × 90; 6, × 600

Globoquadrina himiensis MAIYA, SAITO and SATO, 1976, pp. 410–411, pl. 4, figs. 1a–c, 2a–c.
Neogloboquadrina himiensis MAIYA, SAITO and SATO. — THOMPSON, 1980, pl. 3, fig. 9.

Types: Holotype and Paratype—Sample N-YB-65, lower part of Yabuta Formation, seacliff on eastern coast of Noto Peninsula, about 1 km north of Ozaki, Himi City, Toyama Prefecture, central Japan.

Diagnosis: Test medium to large in size, $4^1/_2$–5 globular chambers in the final whorl, about 14–15 chambers in all, arranged in about 3 whorls. Chambers subspherical to ovate, partially embracing, increasing gradually in size as added, sutures well depressed. Aperture interiomarginal, nearly umbilical-extraumbilical, a low arch with a distinct tooth-like flap. Wall calcareous, non-spinose, uniformly and coarsely pitted.

Remarks: This species is directly descended from *N. kagaensis*, and can be distinguished by its deeper and more open umbilicus, more umbilical-extraumbilical aperture and quadrate outline. It can be separated from *N. eggeri* by having only 5 chambers in the final whorl, whereas *N. eggeri* in the samples examined by the authors had $5^1/_2$–6 chambers, and by its higher trochospire and apertural arch concave towards the periphery.

Distribution: Late Pliocene to Early Pleistocene in the northwest Pacific.

Plate 38 *Neogloboquadrina asanoi* (Maiya, Saito and Sato), 1976
Neogloboquadrina kagaensis (Maiya, Saito and Sato), 1976
Neogloboquadrina himiensis (Maiya, Saito and Sato), 1976

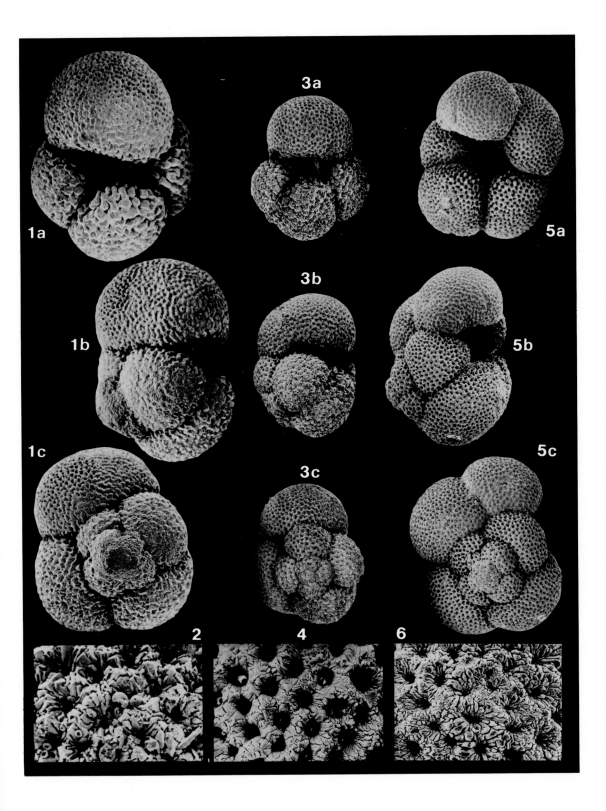

39. 1 *Globorotalia planispira* BRÖNNIMANN and RESIG, 1971
Plate 39, Figs. 1a–c, × 200; 1d, × 1333

Globorotalia (Turborotalia) planispira BRÖNNIMANN and RESIG, 1971, p. 1282, pl. 36, figs. 4,6; pl. 44, figs. 1,2,3(?),4,5,7,8.
Globorotalia planispira BRÖNNIMANN and RESIG. — STAINFORTH *et al.*, 1975, pp. 394–396, fig. 193, nos. 1–6.
Globorotaloides hexagona (NATLAND). — KENNETT (not NATLAND), 1973, pl. 19, figs. 4–6.
Globorotaloides planispira (BRÖNNIMANN and RESIG). — POAG and VALENTINE, 1976, p. 202, pl. 12, figs. 1–9.

Types: Holotype—DSDP Hole 62.1, Core 6, Section 5, 15–17 cm (Lat. 01°52.2′N, Long. 141°56.3′E, 2,607 m).

Plesiotype—DSDP Hole 289, Core 4, core catcher (Lat. 00°29.92′S, Long. 158°30.69′E, 2,206 m).

Diagnosis: Test small, a low trochospire, 5 subglobular chambers in the final whorl in all 12–13 chambers arranged in 2–2¹/₂ whorls. Chambers subspherical becoming slightly flattened radially, size increasing slowly as added, partly embracing, sutures deep. Aperture umbilical-extraumbilical, a low arch beneath a heavy and wide lip. Wall calcareous, finely perforated, coarsely granular.

Remarks: The placement of this taxon into the *Globorotalia* was probably done because of its apertural position and lip; it could equally be placed into *Neogloboquadrina* because of its wall structure and test surface. Most of the specimens illustrated by BRÖNNIMANN and RESIG show only the umbilical view, strongly resembling *N. humerosa* or *N. pachyderma*.

Distribution: BRÖNNIMANN and RESIG (1971) report it from Middle Pliocene (N. 20) to Late Pleistocene (N. 22) in DSDP Hole 62.1.

39. 2 *Globorotalia parkerae* BRÖNNIMANN and RESIG, 1971
Plate 39, Figs. 2a–c, × 333; 2d, × 2000

Globorotalia (Turborotalia) parkerae BRÖNNIMANN and RESIG, 1971, pp. 1280–1281, pl. 43, figs. 7–10; pl. 47, figs. 4,6; pl. 48, figs. 2,3.
Tenuitella parkerae (BRÖNNIMANN and RESIG). — FLEISHER, 1974a, p. 1033.

Types: Holotype—Cap HG 41, 0–1 cm (Lat. 15°56′S, Long 117°14′W, 3,394 m).
Plesiotype—V28-239, 2,099 cm (Lat. 03°15′N, 159°11′E, 3,490 m).

Diagnosis: Test small, a low trochospiral coil, 4¹/₂–5 subglobular chambers in the final whorl, 12–13 chambers arranged in 2–2¹/₂ whorls. Chambers subspherical, initially enlarging slowly in size as added, later enlarging rapidly in a radial direction, loosely embracing, sutures distinct. Aperture umbilical-extraumbilical, small and low, flanked by a prominent flap-like lip. Wall calcareous, thin, finely perforated, slightly pustulate.

Remarks: BRÖNNIMANN and RESIG (1971, p. 1281) called attention to the close ontogenetic similarities between this species and *G. anfracta* (see Pl. 47). *G. parkerae*, however, has less inflated chambers which are more lobulate and radially elongate than *G. anfracta*.

Distribution: BRÖNNIMANN and RESIG (1971) recorded *G. parkerae* from the late Middle Pliocene (N. 20) to Recent in DSDP Hole 62-1.

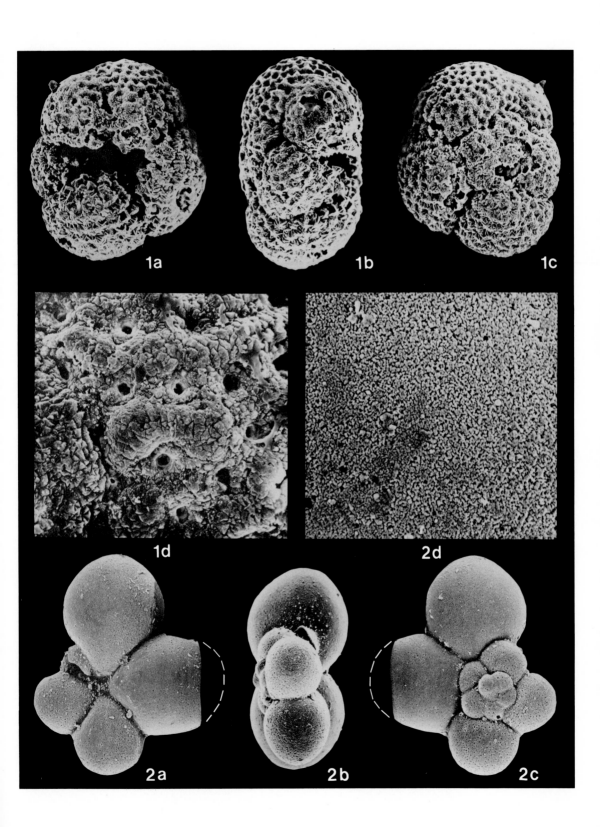

40. 1 *Globorotalia neominutissima* BERMÚDEZ and BOLLI, 1969
Plate 40, Figs. 1a–b, 2a, × 230; 2b, × 2916

Globorotalia neominutissima BERMÚDEZ and BOLLI, 1969, p. 175, pl. 13, figs. 10–12.

Types: Holotype—Sample Caldeon no. 72, collected at La Cantera de Araya, Araya Peninsula, Estado Sucre, Venezuela (Pliocene).

Homotype—V12-98, 1,133 cm (Lat. 10°47.1′N, Long. 65°06.8′W, 736 m).

Diagnosis: Test small, a low to medium height trochospire, 5 chambers in the final whorl; about 11–15 chambers in about $2^1/_2$ whorls. Chambers globular except the last chamber which is ovoid in shape. Chambers increase gradually in size, excepting the last chamber which can be either smaller or considerably larger than the penultimate chamber; sutures simple, well-marked depression. Aperture interiomarginal, umbilical, partially covered by the overhanging final chamber hemmed with a prominent lip. Wall calcareous, coarsely perforated, spinose.

Remarks: The illustration of the holotype specimen (BERMÚDEZ and BOLLI, 1969, pl. 13, figs. 10–12) is quite accurate except for the outline of the pre-antepenultimate chamber. In the umbilical view, this chamber is much more spherical than the illustration which shows the chamber to be elongated toward the umbilicus.

Distribution: From the *Globorotalia menardii* zone to the *Globorotalia truncatulinoides* zone, central and eastern coast of Venezuela.

40. 2 *Globorotalia seigliei* BERMÚDEZ and BOLLI, 1969
Plate 40, Figs. 3a, 4a–b, × 230; 3b, × 2916

Globorotalia seigliei BERMÚDEZ and BOLLI, 1969, pp. 177–178, pl. 18, figs. 1–3.

Types: Holotype—Cariaco Trench, 703 m, Estación de Investigaciones Marinas de Margarita.

Plesiotype—V12-98, 1,133 cm (Lat. 10°47.1′N, Long. 65°06.8′W, 736 m).

Diagnosis: Test small, a low trochospire, 5 chambers in the final whorl; about 15–18 chambers in about $2^1/_2$–3 whorls; chambers gradually enlarging as added; sutures distinct, simple, depressed, slightly curved to radial on both the spiral and umbilical sides; test nearly biconvex, the spiral side is nearly flat but the umbilical side moderately vaulted; aperture interiomarginal, umbilical-extraumbilical, an apertural slit reaching close to the periphery, a narrow lip fringing the apertural opening; wall calcareous, finely pitted.

Remarks: The holotype (USNM 221892) of *G. seigliei* resembles very closely the figured paratype specimen of *Globigerina atlantisae* CIFFELI and SMITH (1970, pl. 1, figs. 3a–c), and it is quite likely that both species are synonymous. The side view of the holotype illustrated by BERMÚDEZ and BOLLI (1969, pl. 18, fig. 2) shows some distortion in the actual shape of the last chamber. The chamber is not quite as spherical as the one shown by them, but is rather elongate at an oblique angle to the peripheral axis.

Distribution: Noted only in Recent sediments from the Cariaco Trench.

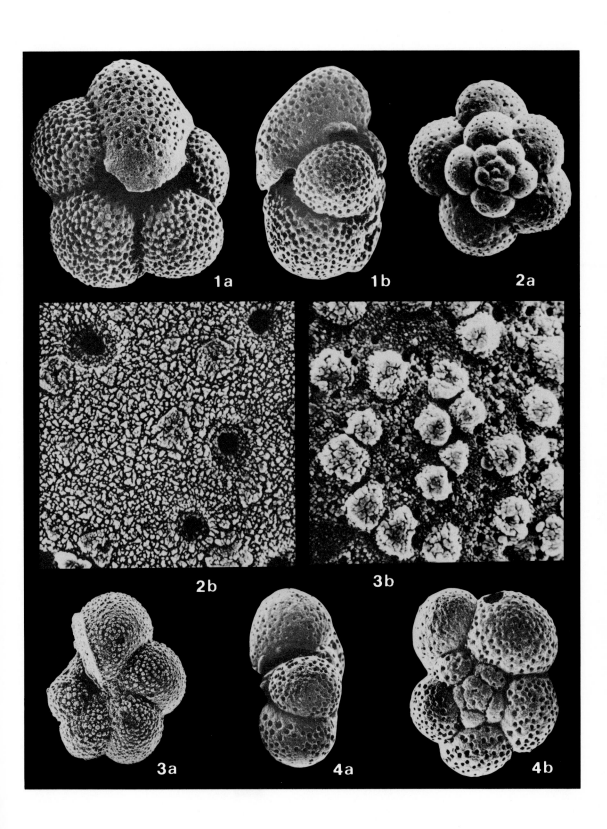

41. 1   *Globorotalia inflata* (D'ORBIGNY), 1839
Plate 41, Figs. 1a–c, × 100; 1d, × 1000
(see also Pl. 56, Fig. 5)

Globigerina trigonula D'ORBIGNY, 1826 (*nomen nudum*), p.277. — FORNASINI, 1903, p. 140, pl. 1, figs. 2, 2a–2b.
Globigerina inflata D'ORBIGNY, 1839b, p. 134, pl. 12, figs, 7–9.
Rotalina aradasii SEGUENZA, 1862, pl. 103, pl. 1, figs. 5, 5a–b.
Globigerina bulloides D'ORBIGNY var. *inflata* D'ORBIGNY. — PARKER and JONES, 1865, pp. 365, 367, pl. 16, figs, 16, 17.
Globigerina nipponica ASANO, 1957, p. 18, pl. 1, figs. 1–3.
Globorotalia trigonula (D'ORBIGNY). — WALLER and POLSKI, pl. 10, figs. 2a–c.
Turborotalia inflata (D'ORBIGNY). — BERMÚDEZ, 1961, pp. 1323–1324, pl. 18, figs. 2a–b.
Globorotalia inflata (D'ORBIGNY). — PARKER, 1962, p. 236, pl. 5, figs. 6–9.
Globorotalia (Turborotalia) inflata (D'ORBIGNY). — BANNER and BLOW, 1967, pp. 145–146, pl. 4, figs. 1a–c (neotype), 11.

Types: Holotype (*inflata*)—Santa, Cruz, TÉNÉRIFFE, Canary Islands (Lat. 28°20′N, Long. 16°30′W).

Syntypes (*aradasii*)—Aci-Castello, dintórni di Catania (Pleistocene); Messinian (Pliocene), Sicily.

Neotype—Challenger Station 8 (Lat. 23°03′15″N, Long. 07°27′00″W, 620 fm).

Holotype (*nipponica*)—Soyo-maru Station 355, Kii Channel, between eastern coast of Shikoku Island and southern coast of Honshu Island, Japan, 439 m.

Plesiotype—V23-95, trigger core top (Lat. 30°24′N, Long. 18°23′W, 4,598 m).

Diagnosis: Test size variable, low trochospire, 4 subglobular chambers in the final whorl, about 14 chambers arranged in 3 whorls. Chambers more inflated on umbilical side than on spiral side, increasing moderately in size as added, partly embracing, sutures distinct but not deep. Aperture a high, wide arch, interiomarginal, intraumbilical-extraumbilical with no apparent lip. Wall calcareous, thick, smooth, very finely perforated, non-spinose although coarsely granular in the umbilicus.

Remarks: The work of BANNER and BLOW (1967) clearly brought out the distinction not previously made between *G. inflata* and *Pulleniatina* spp., in terms of coiling, chamber morphology and apertural location. *G. inflata* may be distinguished from *G. oscitans* by the usually larger size and larger aperture without lip.

Distribution: Middle or late Miocene (N. 16) to Recent, cool-temperate waters.

41. 2   *Globorotalia oscitans* TODD, 1958
Plate 41, Figs. 2a–c, × 100; 2d, × 1000

Globorotalia oscitans TODD, 1958, p. 201, pl. 1, figs. 23a–c.

Types: Holotype—Atlantis sample 5Y, Core 11, 529–530 cm (Lat. 38°52′N, Long. 10°35′E, 2,444 m).

Plesiotype—V10-72, trigger core top (Lat. 39°37′N, Long, 11°54′E, 3,125 m).

Diagnosis: Test small, very low trochospire, 3–4 chambers visible in the final whorl, about 10 chambers in all arranged in about $2\frac{1}{2}$ whorls. Chambers subglobular, more

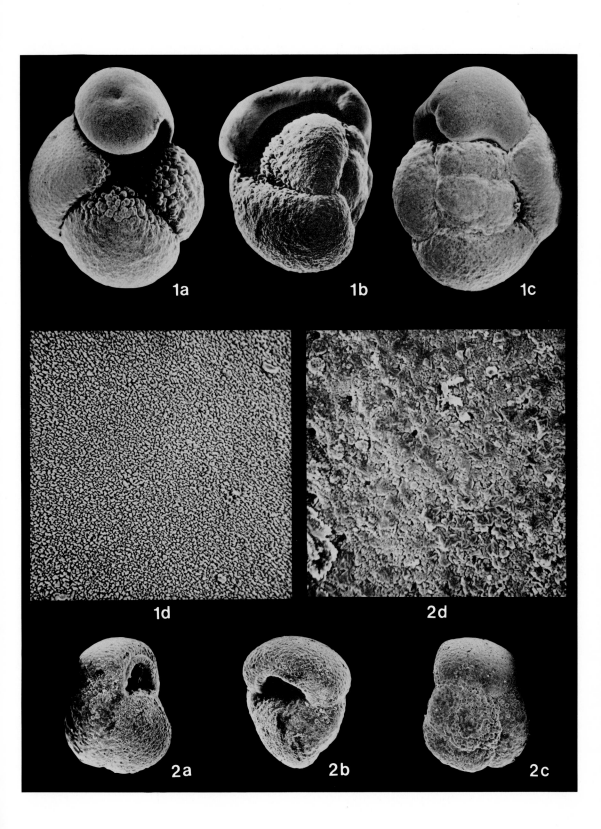

inflated ventrally than dorsally, giving the test a flattened spiral side, chamber size increasing moderately as added, sutures weak. Aperture a large, rounded opening surrounded by a prominent lip. Wall calcareous, thick, finely perforated, smooth to pustulate.

Remarks: TODD (1958) distinguished this species from *G. inflata* by its having "fine blunt spines," which are here interpreted as pustules. Juvenile forms referable to *G. inflata*, however, have also been noted to be pustulate (THOMPSON, 1973, p. 470). *G. oscitans* is here distinguished from *G. inflata* by its narrower and more symmetrical aperture and thin but prominent apertural lip.

Distribution: Pleistocene? to Recent, reported only from the Mediterranean.

42. 1 *Globorotalia triangula* THEYER, 1973
Plate 42, Figs. 1a–c, × 160; 1d, × 1600

Globorotalia inflata (D'ORBIGNY) (variant) — LAMB and BEARD, 1972, p. 53, pl. 27, figs. 1,2,5.
Globorotalia inflata triangula THEYER, 1973, pp. 199–201, pl. 1, figs. 1–7.
Globorotalia triangula THEYER. — STAINFORTH, LAMB, LUTERBACHER, BEARD, and JEFFORDS, 1975, p. 416, fig. 208, nos. 1–7.

Types: Holotype—Eltanin 39-76, 955–960cm (Lat. 36° 30′S, Long. 161° 41′E, 3,785 m).
Plesiotype—V18-227, 435 cm (Lat. 39° 26′S, Long. 165° 41′E, 3,052 m).

Diagnosis: Test medium in size, 3 rhomboid chambers in the final whorl, about 8–10 chambers in all arranged in a very low trochospire of about 2 whorls. Chambers, from the spiral side, lunate to subquadrate, flat, overlapping, sutures curved, weakly depressed; from the ventral side, subtriangular, strongly inflated, sutures nearly radial. Aperture interiomarginal, umbilical to umbilical-extraumbilical, a very low arch without lip. Wall calcareous, very finely perforated, non-spinose, weakly pustulate on the ventral surface.

Remarks: THEYER (1973) separated this subspecies from *G. inflata s.s.* by its highly vaulted ventral side, flat dorsal side, narrow aperture and maximum of 3 chambers in the final whorl.

Distribution: Reported only from the latest Pliocene and middle Pleistocene from the Tasman Basin.

42. 2 *Globorotalia hessi* BOLLI and PREMOLI-SILVA, 1973
Plate 42, Figs. 2a–c, × 80; 2d, × 1600

Globorotalia crassaformis B, BOLLI, 1970, pp. 580–581, pl. 4, figs. 13–16.
Globorotalia crassaformis (GALLOWAY and WISSLER). — LAMB and BEARD, 1972 pl. 21, figs. 1–8.
Globorotalia hessi BOLLI and PREMOLI-SILVA, 1973, pp. 476–477 (no figures).

Types: Holotype—DSDP Leg. 4, Site 29, Core 1, Section 3, 10–12 cm (Lat. 14° 47.11′N, Long. 69° 19.36′W, 4,247m).
Near-Topotype—V26-120, 1,025 cm (Lat. 14° 27′N, Long. 68° 33′W, 4,707 m).

Diagnosis: Test medium to large in size, 4 rhomboid chambers in the final whorl, about 10 chambers in all arranged in about $2^1/_2$ whorls in a very low trochospire with a weak peripheral keel. Chambers, from the spiral side, subquadrate, partially overlapping, weakly inflated, sutures curved, depressed; from the umbilical side, subtriangular, partially overlapping, strongly inflated, sutures slightly curved, depressed. Aperture interiomarginal, umbilical-extraumbilical, a low slit at the base of the apertural face of the final chamber and showing a rim-like lip. Wall calcareous, densely perforated in later chambers, usually heavily overgrown with cortex on early whorls, non-spinose, coarsely granular.

Remarks: BOLLI and PREMOLI-SILVA (1973) distinguished this species from *G. crassaformis* by its larger, more robust size, more quadrangular outline, less lobate equatorial profile and typically concave spiral side.

Distribution: Restricted to Early-Middle Pleistocene, reported only from the Caribbean.

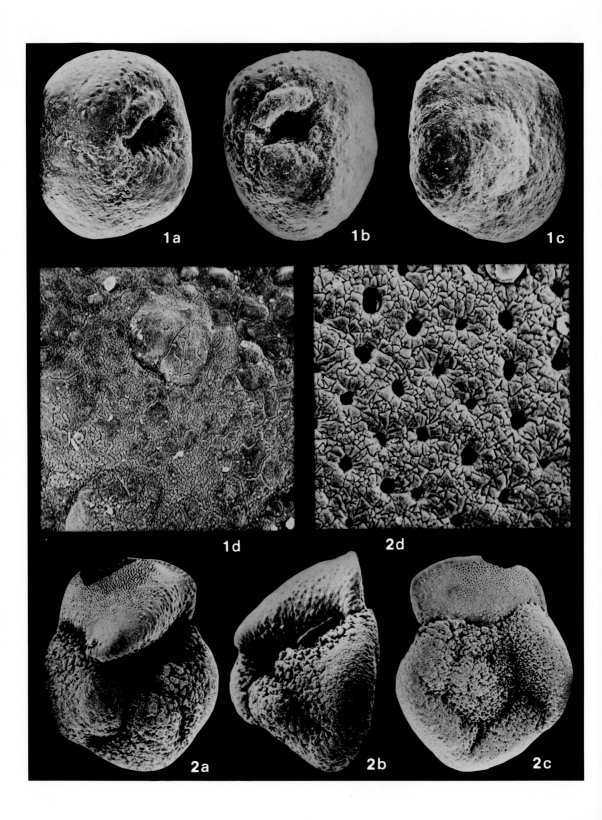

43. 1   *Globorotalia puncticulata* (DESHAYES), 1832
Plate 43, Figs. 1a–c, × 120; 1d, × 800

Globigerina punctulata D'ORBIGNY, 1826, p. 277, No. 8 (*nomen nudum*). — FORNASINI, 1899, p. 210,
text-fig. 5 (after D'ORBIGNY). — BANNER and BLOW, 1960a, p. 15 (lectotype).
Globigerina puncticulata "D'ORBIGNY," nobis, DESHAYES, 1832, p. 170 (no figures). — BANNER and
BLOW, 1960a, p. 15, pl. 5, figs. 7a–c (lectotype).
Globorotalia puncticulata (DESHAYES). — KANE, 1953, p. 30, pl. 1, fig. 9.

Types: Syntype (*puncticulata*)—From Rimini, northeastern Italy, on the Adriatic Sea.
Lectotype (*puncticulata*)—from the original materials of D'ORBIGNY.
Plesiotype—Trubi Marl, roadside outcrop between Agrigento and Porto
Empedocle, southern coast of Sicily, collected by W.A. BERGGREN.

Diagnosis: Test medium to large in size, 4 rhomboid chambers in the final whorl,
about 12 chambers in all arranged in a low trochospine of about $2^1/_2$ whorls. Chambers,
from the spiral side, subquadrate and radially compressed; weakly inflated, partially
embracing, sutures depressed; from the umbilical side, subtriangular, much inflated,
sutures depressed. Aperture interiomarginal, umbilical-extraumbilical, a high arch
lacking a rim or lip. Wall calcareous, densely perforated, non-spinose, heavily punctate.

Remarks: BANNER and BLOW (1960a) have ably covered the nomenclatural difficulties
of this species. It can be distinguished from the *G. crassaformis* group by its lack of
keel and lobulate periphery. PARKER (1967, p. 179) suggested that it might be a variant
form of *G. inflata* differing in the extent of pustules and lobulation, but this relationship
has not been confirmed.

Distribution: Middle Pliocene (N. 20) to Recent, subtropical and temperate.

43. 2   *Globorotalia crassaformis* (GALLOWAY and WISSLER), 1927
Plate 43, Figs. 2a–c, × 120; 2d, × 800

Pulvinulina crassa BRADY (not *Rotalina crassa* D'ORBIGNY, 1840), 1884, p. 694, pl. 103, figs. 11,12.
Globigerina crassaformis GALLOWAY and WISSLER, 1927, p. 41, pl. 7, fig. 12.
Globorotalia (Pulvinulina) crassa (D'ORBIGNY). — HERON-ALLEN and EARLAND, 1932, p. 428.
Globorotalia crassa (D'ORBIGNY). — CUSHMAN, TODD and POST, 1954, p. 370, pl. 91, fig. 16.
Globorotalia crassaformis (GALLOWAY and WISSLER). — PARKER, 1962, p. 235, pl. 4, figs. 17,18,20,21.
Globorotalia (Turborotalia) crassaformis crassaformis (GALLOWAY and WISSLER). — BLOW, 1969, p. 347,
pl. 4, figs. 1–3; pl. 37, figs. 1–4.
Globorotalia crassaformis crassaformis (GALLOWAY and WISSLER). — RÖGL and BOLLI, 1973, p. 568,
pl. 7, figs. 7–14; pl. 16, figs. 7,8.
Globorotalia (Truncorotalia) crassaformis (GALLOWAY and WISSLER). — FLEISHERA, 1974, p. 1028.

Types: Holotype—Middle bed in the Lomita Quarry, Los Angeles, California.
Topotype—Sample collected by A.R. LOEBLICH, JR.

Diagnosis: Test small to medium in size, 4 rhomboid chambers in the final whorl,
about 12 chambers in all, arranged in a very low trochospire of $2^1/_2$ whorls, with a prominent peripheral keel. Chambers, from the spiral side, ovate to subquadrate, radially
compressed, not inflated, partially overlapping, sutures slightly depressed from the
umbilical side, subtriangular, strongly inflated, sutures depressed. Aperture umbilical-
extraumibilical, an interiomarginal circle at the base of the umbilical face, sometimes

showing a faint rim-like lip. Wall calcareous, finely perforated, non-spinose, covered with irregularly spaced pustules.

Remarks: Many specimens have been referred to this species, many of which with rounded peripheries with little or no keel, would better be referred to *G. hessi, G. oceanica* or *G. ronda*. On the holotype as well as topotypic material, the peripheral keel illustrated by GALLOWAY and WISSLER is plainly visible and formed much of their concept, and served as the basis for CUSHMAN and BERMÚDEZ (1949) to later separate *G. oceanica*.

Distribution: Late Miocene (N. 16) to Recent, subtropical to temperate.

Plate 43 *Globorotalia puncticulata* (Deshayes), 1832
Globorotalia crassaformis (Galloway and Wissler), 1927

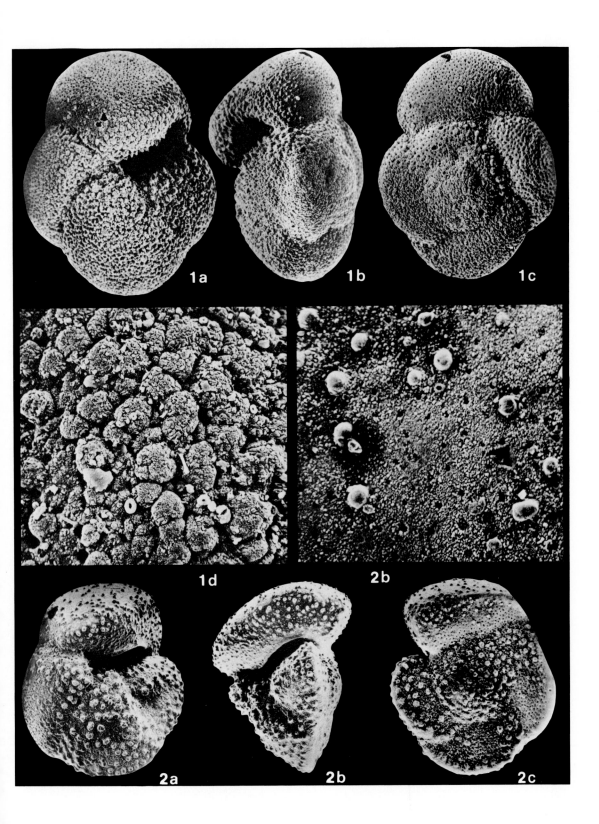

44. 1   *Globorotalia oceanica* CUSHMAN and BERMÚDEZ, 1949
Plate 44, Figs. 1a–c, × 100; 1d, × 1000

Pulvinulina crassa (D'ORBIGNY).   —   BRADY, 1884 (not *Rotalina crassa* D'ORBIGNY, 1840), pl. 103, fig. 12
(not fig. 11 = *G. crassaformis*).   —   CUSHMAN, 1915, p. 58, pl. 27, fig. 1.
Globorotalia oceanica CUSHMAN and BERMÚDEZ, 1949, pp. 43–44, pl. 8, figs. 13–15.
Globorotalia (Turborotalia) crassaformis oceanica CUSHMAN and BERMÚDEZ.   —   BLOW, 1969, p. 348,
pl. 4, figs. 7–9; pl. 37, fig. 5.
Globorotalia crassaformis oceanica CUSHMAN and BERMÚDEZ.   —   RÖGL, 1974, p. 749, text-fig. 4, nos. 7–
12; pl. 3, figs. 19–21.

Types: Holotype—Atlantis Station 2980 (Lat. 22°47′N, Long. 78°40′W, 250 fm).
Plesiotype—V31-9, 80 cm (Lat. 28°12′S, Long. 31°29′W, 4,102 m).

Diagnosis: Test medium in size, 4 rhomboid chambers in the final whorl, about 10 chambers in all arranged in a very low trochospire of about 2 whorls. Chambers, from the spiral side, subquadrate, weakly inflated, radially compressed, slightly embracing, sutures nearly radial, slightly incised; from the umbilical side, they are triangular, strongly inflated, sutures radial, well incised. Aperture interiomarginal, umbilical-extraumbilical, a low, slit-like opening overhung by a thin rim-like lip. Wall calcareous, perforated, non-spinose, coarsely pustulate especially on the earlier whorls.

Remarks: BLOW (1969) differentiated *G. oceanica* from *G. crassaformis s.s.* by its rounded peripheral shoulders, more inflated dorsal surface chambers, open and deep umbilicus and less conical ventral side. *G. oceanica* was distinguished from *G. ronda* as having less embracing chambers, a loose coil and a larger umbilicus.

Distribution: Late Miocene (N. 17) to Recent, subtropical to temperate.

44. 2   *Globorotalia ronda* BLOW, 1967
Plate 44, Figs. 2a, 3a–b × 100; 2b, × 1000

Globorotalia (Turborotalia) crassaformis ronda BLOW, 1969, pp. 388–390, pl. 4, figs. 4–6; pl. 37, figs. 6–9.
Globorotalia crassaformis ronda BLOW,   —   RÖGL, 1974, p. 749, fig. 4, nos. 13–15; pl. 3, figs. 7–18.

Types: Holotype—WHB sample 181A, Bowden Formation, Buff Bay beds, Jamaica.
Plesiotype—V3-144 B, core top (Lat. 24°13′N, Long 81°52′W, 589 m).

Diagnosis: Test medium to large in size, 4 rhomboid chambers in the final whorl, 14–15 chambers in all arranged in a low trochospire of about 2–2$^1/_2$ whorls, periphery slightly thickened but no true keel. Chambers, from the spiral side, subquadrate, slightly inflated, closely packed, size increases slowly as added, sutures slightly depressed; from the ventral side, triangular, strongly inflated, sutures depressed, peripheral outline broadly rounded. Aperture interiomarginal, umbilical-extraumbilical, a low arch bordered by a thin lip. Wall calcareous, perforated, non-spinose, coarsely pustulate on early chambers.

Remarks: BLOW (1969) distinguished this form from *G. oceanica* by its more closely packed chamber, smaller aperture and thick pustulate crust, and from *G. crassaformis* by its rounded periphery, tighter chamber packing, less conical ventral side and more radially compressed chambers.

Distribution: Late Pliocene (N. 17) to early Pleistocene (N. 22), subtropical to temperate.

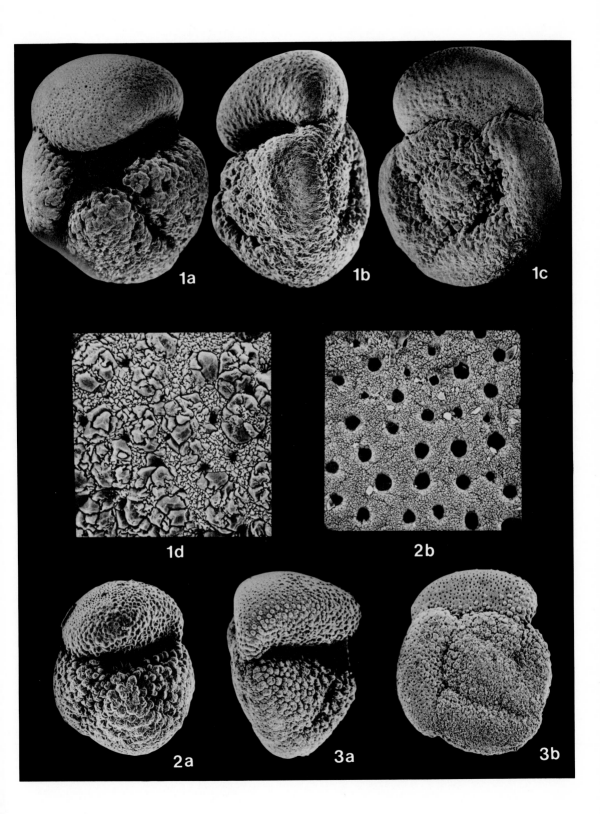

45. 1 *Globorotalia crassula* CUSHMAN and STEWART, 1930
Plate 45, Figs. 1a, 2a–b, × 100; 1b, × 1000

Globorotalia crassula CUSHMAN and STEWART, 1930, p. 77, pl. 7, figs. ? 1a, 1b, 1c. — PARKER, 1967, p. 177, pl. 29, figs. 2,7.

Globorotalia punctulata (D'ORBIGNY). — PHLEGER and PARKER, 1951, p. 36, pl. 20, figs. 3–7.

Globorotalia hirsuta PARKER (part) (not *Rotalina hirsuta* D'ORBIGNY, 1839), 1962, p. 236, pl. 5, figs. 13,15 (not pl. 5, figs. 10–12; pl. 6, fig. 1)

cf. *Globorotalia hirsuta aemiliana* COLALONGO and SARTONI, 1967, pp. 267–274, pl. 30, figs. 1–5; pl. 31, figs. 2–4, tf. 2.

Globorotalia (Globorotalia) crassula crassula CUSHMAN and STEWART. — BLOW, 1969, pp. 361–362, pl. 9, figs. 1–3 (holotype refigured).

Globorotalia crotonensis CONATO and FOLLADOR, 1967, pp. 556–557, tf. 1, no. 1; tf. 4, nos. 1–2. — BÉ, 1977, pl. 11, figs. 31a–c.

Types: Holotype—Station 7, diggings for spring near house above North Fork of Elk River about 30 yards west of center of Sec. 30, T.4N., R.1E., Humboldt County, California.

Plesiotype—V24-101, trigger core top (Lat. 13°10′N, Long. 178°53′E, 3,336 m).

Diagnosis: Test medium in size, 4 rhomboid chambers in the final whorl, about 10 chambers in all arranged in a low trochospire of about 2 whorls; a wide peripheral thickening suggests the presence of a keel. Chambers, from the spiral side, subquadrate to lunate, very weakly inflated, weakly-incised curving sutures; from the ventral side they are subtriangular to nearly subquadrate, strongly inflated, incised sutures. Aperture interiomarginal, umbilical-extraumbilical, a low arch overhung with a rim-like lip. Wall calcareous, finely perforated, non-spinose, pustulate, especially on the early chambers of the initial whorl on the ventral side.

Remarks: BLOW (1969) distinguished *G. crassula* from *G. viola* as having more tightly packed chambers and higher ventral vaulting. It differs from members of the *G. crassaformis* group by being less compressed radially, and from the *G. hirsuta* group by having more inflated chambers and fewer pustules.

Distribution: Middle Pliocene (N. 18) to Recent, temperate waters.

45. 2 *Globorotalia viola* BLOW, 1969
Plate 45, Figs. 3a, 4a–b, × 100; 3b, × 1000

Globorotalia (Globorotalia) crassula viola BLOW, 1969, pp. 397–398, pl. 5, figs. 4–9.

Types: Holotype—Sample WHB 181B, Bowden Formation, Buff Bay beds, Jamaica. Topotype—Same sample.

Diagnosis: Test large in size, 4 rhomboid chambers in the final whorl, 12 chambers in all arranged in a low trochospire of about $2^1/_2$ whorls, with a distinct peripheral keel. Chambers, from the spiral side, lunate to subquadrate, nearly flat, with slightly depressed, curved, limbate sutures, slightly thickened with carina material; from the ventral side, they are subtriangular, moderately inflated, with well-depressed sutures; chambers increase slowly in size as added. Aperture interiomarginal, umbilical-extraumbilical, a low arch with a thin rim-like lip. Wall calcareous, finely perforated, pores without pore pits, non-spinose, sporadically pustulate.

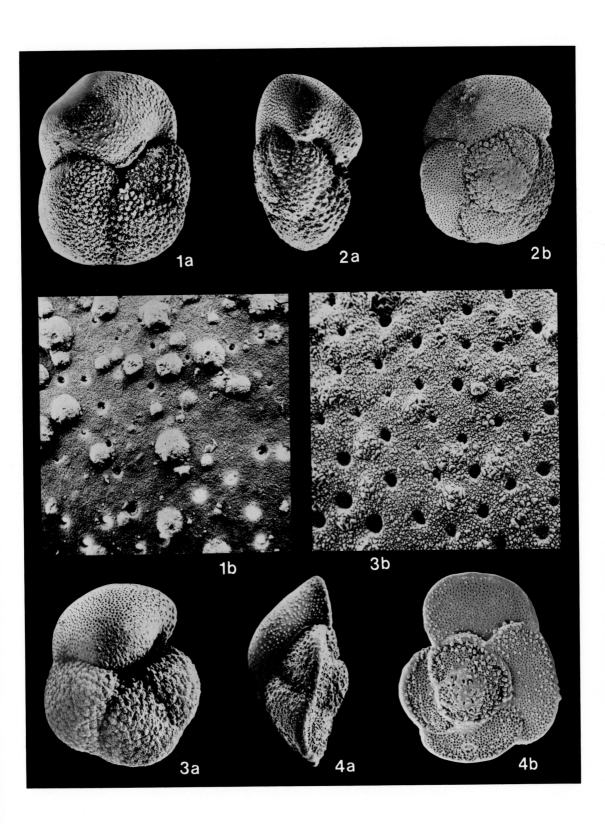

Remarks: This species can be distinguished from *G. crassula* by its less inflated chambers, less compact chamber arrangement and sharply acute periphery with distinct keel, and from the *G. hirsuta* group by the lower trochospire and less radially compressed chambers.

Distribution: Middle Pliocene (N. 18) to Recent, more equatorial than *G. crassula*.

46.1 *Globorotalia hirsuta* (D'ORBIGNY), 1839
Plate 46, Figs. 1a–c, × 71; 1d, × 714

Rotalina hirsuta D'ORBIGNY, 1839b, p. 131, pl. 1, figs. 37–39.
Rotalina canariensis D'ORBIGNY, 1839b, p. 130, pl. 1, figs. 34–36. — BANNER and BLOW, 1960a,
  pp. 33–34, pl. 5, fig. 4 (*nomen dubium*) (lectotype).
Pulvinulina elegans (D'ORBIGNY) var. *hirsuta* (D'ORBIGNY). — JONES and PARKER, 1872, p. 114.
Pulvinulina canariensis (D'ORBIGNY). — BRADY, 1884, p. 692, pl. 103, figs. 8a–10b.
Globorotalia hirsuta (D'ORBIGNY). — CUSHMAN, 1931, p. 99, pl. 17, fig. 6.
Globorotalia (Globorotalia) hirsuta hirsuta (D'ORBIGNY). — BLOW, 1969, pp. 398–400, pl. 8, figs. 1–3;
  pl. 43, figs. 1–2 (neotype).
Globorotalia (Hirsutella) hirsuta (D'ORBIGNY). — BANDY, 1972, pp. 298, 310.
Globorotalia (Hirsutella) hirsuta hirsuta (D'ORBIGNY). — FLEISHER, 1974a, p. 1027.
Globoquadrina patriciae MCCULLOCH, 1979, p. 417, pl. 193, fig. 9.

Types: Syntypes (*canariensis*)—Near the Canary Islands (Lat. 28°29′N, Long. 16°30′
        W).
        Neotype (*hirsuta*)—Challenger Station 8, off Gomera, Canary Islands (Lat.
        23°12′N, Long. 32°56′W, 2,700 fm).
        Near-topotype—V23-95, trigger core top (Lat. 30°24.2′N, Long. 18°23′W,
        4,598 m).

Diagnosis: Test large in size, 4 rhomboid chambers in the final whorl, 9–10 chambers
in all arranged in a medium-height biconvex trochospire of about $2^1/_2$ whorls, with a
distinct keel. Chambers, from the spiral side, lunate, flat to very weakly inflated, partially
overlapping, sutures curved, depressed; from the umbilical side, they are rounded-triangular, moderately inflated, with distinct incised sutures. Aperture umbilical-extraumbilical, low interiomarginal arch distorted by an inflated swelling of the final chamber
and overhung by a thin lip. Wall calcareous, finely perforated, non-spinose, covered
with unconnected coarse pustules.

Remarks: This species is characterized by its relatively high spire, pustulate surface,
weakly lobulate periphery and uninflated dorsal surface.

Distribution: Pleistocene (N. 22) to Recent, temperate waters.

46.2 *Globorotalia scitula* (BRADY), 1882
Plate 46, Figs. 2a–c, × 71; 2d, × 714

Pulvinulina scitula BRADY, 1882, pp. 716–717 (no figures). — BANNER and BLOW, 1960a, pp. 27–29, pl. 5,
  figs. 5a–c (lectotype).
Pulvinulina patagonica (D'ORBIGNY), BRADY (not *Rotalina patagonica* D'ORBIGNY, 1839), 1884, pl. 103, fig. 7.
Globorotalia patagonica (D'ORBIGNY). — WIESNER, 1931, p. 135.
Globorotalia scitula (BRADY). — CUSHMAN and HENBEST, 1940, p. 36, pl. 8, figs. 5a–c. — CUSHMAN,
  1927, p. 175. — HERB, 1968, p. 477, pl. 2, figs. 4a–c.
Globorotalia scitula scitula (BRADY). — BLOW, 1959, p. 219, pl. 19, figs. 120a–c.
Globorotalia (Turborotalia) scitula (BRADY). — BANNER and BLOW, 1960, pp. 27, 29, pl. 5, fig. 5.
Globorotalia (Hirsutella) scitula scitula (BRADY). — FLEISHER, 1974a, p. 1028.

Types: Holotype—Knight Errant Expedition Station 7, Faröe Channel (Lat. 59°37′N,
        Long. 07°19′W, 530 fm).
        Lectotype—From original materials.
        Plesiotype—V28-294, core top (Lat. 28°26′N, Long. 139°58′E, 2,308 m).

Diagnosis: Test medium to large in size, 4–6 wedge-shaped chambers in the final whorl, about 15 chambers in all arranged in a low biconvex trochospire of about 3 whorls, with no evidence of a peripheral keel. Chambers from the spiral side lunate or compressed, semi-circular, weakly inflated, partially embracing, size increasing slowly as added, giving the test a weakly lobulate, nearly circular peripheral outline, sutures curved, depressed, chambers often imbricated in the final whorl; from the ventral side, they are subtriangular, moderately inflated, sutures slightly curved, depressed. Aperture interiomarginal, umbilical-extraumbilical, a low, asymmetrical arch with a heavy rim-like lip. Wall calcareous, uniformly perforated by well-spaced, large pores with wide pore pits on earlier chambers; non-spinose, lightly pustulate near the aperture in the ventral side.

Remarks: The lectotype described by BANNER and BLOW (1960a) is quite small whereas most specimens observed from the Pacific grow to fairly large size. *G. scitula* can be distinguished from *G. hirsuta* by its smooth surface and subcircular equatorial profile.

Distribution: Early Miocene (N. 9) to Recent, equatorial to cool temperate.

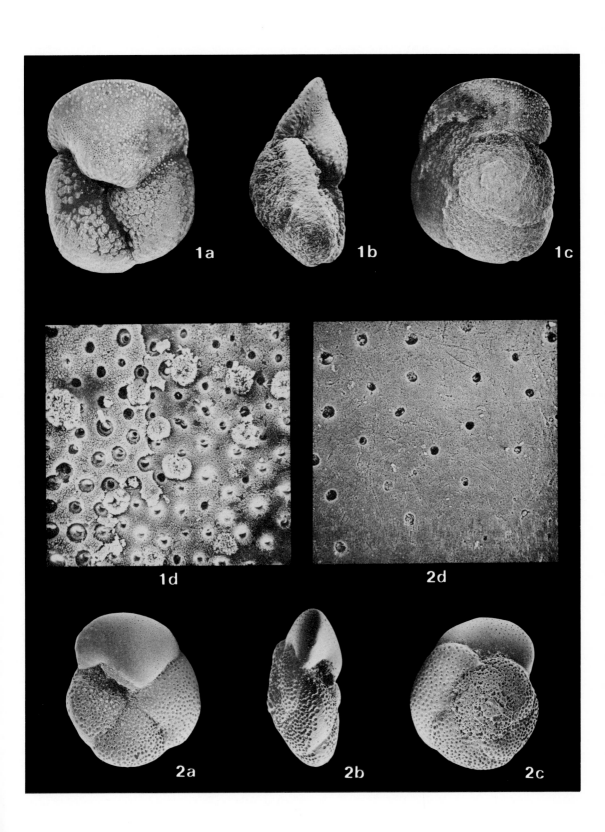

47. 1   *Globorotalia theyeri* FLEISHER, 1970
Plate 47, Figs. 1a–c, × 90; 1d, × 1430

Globorotalia hirsuta (D'ORBIGNY).   —   PARKER, 1962, pl. 5, figs. 12,14.
Globorotalia (Hirsutella) theyeri FLEISHER, 1974a, p. 1028, pl. 12, fig. 9; pl. 13, figs. 1–5.

Types: Holotype—DSDP Leg 23, Hole 219, Core 1, Section 1, 110–112 cm. (Lat.
09°01.75′N, Long. 72°52. 67′E, 1,764 m).

Topotype—DSDP Leg 23, Hole 219, Core 1, Section 4, 70–72 cm, donated
by R. L. FLEISHER.

Diagnosis: Test large in size, 4–5 rhomboid chambers in the final whorl, about 13–16
chambers in all arranged in a low trochospire of about $2^1/_2$ whorls, often showing a very
slight thin keel. Chambers from the spiral side lunate but not greatly compressed radially, giving a well-lobulated equatorial profile, very weakly inflated, sutures well depressed. Aperture interiomarginal, umbilical-extraumbilical, a moderate arch with a
small rim-like lip. Wall calcareous, coarsely perforated, non-spinose, weakly pustulate
on the ventral side near the umbilicus.

Remarks: This species can be distinguished from *G. scitula* and *G. crozetensis* by its
more lobate periphery, and from the *G. hirsuta* group by its lower trochospiral coil and
smooth surface.

Distribution: FLEISHER (1974a) recorded it from middle Pleistocene (N. 22) to Recent
in the Arabian Sea and referred to several examples in the tropical Indian Ocean.

47. 2   *Globorotalia crozetensis* THOMPSON, 1973
Plate 47, Figs. 2a–c, × 90; 2d, × 1430

Globorotalia crozetensis THOMPSON, 1972, p. 471, pl. 2, figs. 1–6.

Types: Holotype—RC8-39, 60 cm (Lat. 42°53′S, Long. 42°21′E, 4,330 m); holotype
refigured.

Diagnosis: Test large in size, $4^1/_2$–5 rhomboid chambers in the final whorl, 14–17
chambers in all arranged in a low trochospire of about 3 whorls, a weak peripheral keel
often present. Chambers from the spiral side weakly lunate, partially overlapping, often
imbricated slightly in the final whorl, weakly inflated sutures slightly depressed; from the
ventral side, they are rounded-triangular, strongly inflated, sutures weakly depressed
and slightly curved, lobulation less than on the spiral side. Aperture interiomarginal,
umbilical-extraumbilical, a low slit-like arch under an often quite loose, plate-like umbilical lip. Wall calcareous, very finely perforated and shiny, non-spinose, weakly
pustulate on the ventral side near the umbilicus.

Remarks: This form can be distinguished from the *G. truncatulinoides* group by its
more inflated spiral side and weak keel. It has not been reported since the initial occurrence and may well be a local species.

Distribution: THOMPSON (1973) reported this species from a late Pleistocene core in
the temperate waters of the Indian Ocean.

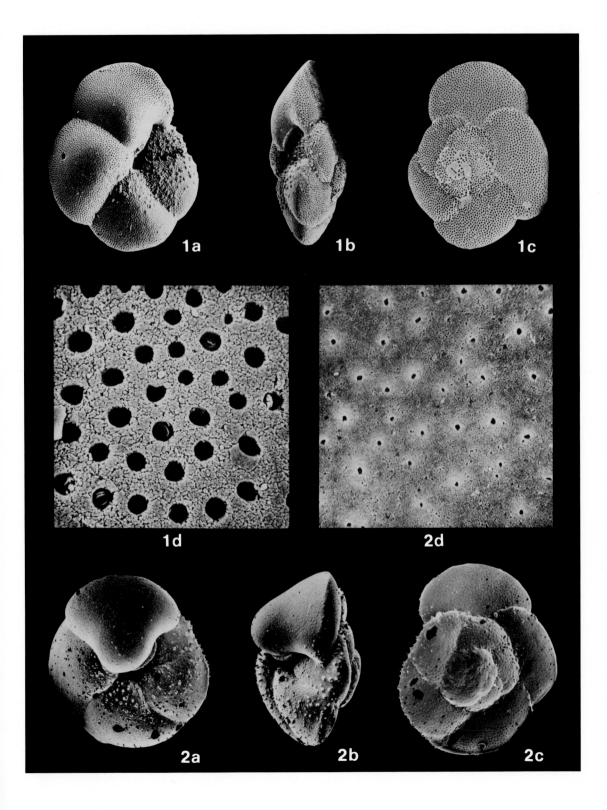

48. 1   *Globorotalia wilesi* THOMPSON, 1980
Plate 48, Figs. 1a–c, × 200; 1d, × 1666

Globorotalia sp. PARKER, 1964, p. 631, pl. 102, figs. 20–22.
Globorotalia sp. RÖGL and BOLLI, 1973, p. 569, pl. 7, figs. 1–4; pl. 16, fig. 4.
Globorotalia wilesi THOMPSON, 1980, p. 785, pl. 4, figs. 1–3.
Globorotalia scitula BRADY small variety. — KELLER, 1978, pp. 101,103, pl. 4, figs. 9–11.

Types: Holotype—DSDP Hole 435, Core 12, Section 1, 65–67 cm (Lat. 39° 44.09′N,
Long. 143°47. 53′E, 3,401 m).
Plesiotype—RC10-161, 60 cm (Lat. 33°05′N, Long. 158° 00′E, 3,587 m).

Diagnosis: Test free, small, biconvex, about 14 subglobular chambers arranged in a
low trochospire of about 3 whorls, with $4^1/_2$–5 conical chambers in the final whorl.
Chambers increasing gradually in size as added, slightly inflated on the spiral side,
moderately inflated on umbilical side, equatorial periphery slightly lobulate, sutures
slightly curved. Aperture interiomarginal, umbilical-extraumbilical, a symmetrical arch
of moderate height at the base of the umbilical shoulder, bordered by a thick umbilical
lip. Wall calcareous, pustulate, coarsely perforated with large pores mostly on the
spiral side.

Remarks: Many workers have probably overlooked hist species as merely a small
variety of *G. scitula*, although PARKER (1964) felt that it probably was a separate taxon.
It has a different distribution in the North Pacific than *G. scitula*.

Distribution: Temperate North Pacific, southern Indian Ocean, (?) Caribbean.
Observed at DSDP Site 435 from Early Pliocene to Recent.

48. 2   *Globorotalia eastropacia* BOLTOVSKOY, 1974
Plate 48, Figs. 2a, 3a–b, × 200; 2b, × 1666

Globorotalia hirsuta eastropacia BOLTOVSKOY, 1974, pp. 127–133, pl. 1, figs. 1–11.
Globorotalia hirsuta (D'ORBIGNY). — BRADSHAW, 1959, p. 44, pl. 8, figs. 1–2. — PARKER, 1962,
p. 238, pl. 5, figs. 13–15. — BANERJI, SCHAFER and VINE, 1971, pl. 6, figs. 1–3.

Types: Holotype—R/V Undaunted Station 163, 228 m plankton tow (Lat. 02°59′N,
Long. 92°00′W).
Plesiotype—V21-210, core top (Lat.02°44′N, Long.90°19′W, 2,140 m).

Diagnosis: Test small to medium in size, 4–5 wedge-shaped chambers in the final
whorl, 13–16 chambers in all arranged in a low biconvex trochospire of about $2^1/_2$
whorls, the periphery is slightly imperforate but with no true keel. Chambers from the
spiral side reniform, flat to slightly inflated, partially embracing, equatorial periphery
distinctly lobulate, size increases moderately as added, sutures curved, depressed, chambers often imbricated; from the ventral side, they are subtriangular, moderately inflated,
slightly overlapping, sutures slightly curved, well depressed. Aperture interiomarginal,
umbilical-extraumbilical, a long, low arch covered by a thin, rim-like lip. Wall calcareous, finely perforated, slightly pustulate, mostly near the aperture on the ventral side.

Remarks: The species was distinguished from *G. hirsuta* by BOLTOVSKOY (1974) by
being more vaulted ventrally, slightly smaller, thin-walled, less pustulate, having fewer
chambers and lacking a keel. In general, it is characterized by its lobulate periphery and
thin, smooth walls.

Plate 48 *Globorotalia wilesi* THOMPSON, 1980
Globorotalia eastropacia BOLTOVSKOY, 1974
Globorotalia anfracta PARKER, 1967

Distribution: The described specimens were collected from plankton tows in waters less than 250 m deep in the eastern equatorial Pacific.

48. 3 *Globorotalia anfracta* PARKER, 1967
Plate 48, Figs. 4a–c, × 200; 4d, × 1666

Globorotalia anfracta PARKER, 1967, p. 175, pl. 28, figs. 3–8.
Globorotalia (Turborotalia) anfracta PARKER. — BRÖNNIMANN and RESIG, 1971, pl. 43, figs. 2,3,6; pl. 48, fig. 4(?).
Turborotalita anfracta (PARKER). — RÖGL and BOLLI, 1973, p. 571, pl. 8, figs. 1–7; pl. 15, fig. 10.
Tenuitella anfracta (PARKER). — FLEISHER, 1974a, p. 1033, pl. 17, figs. 9,10.

Types: Holotype—CHUB IV G, 0–7.5 cm (Lat. 13°10′N, Long. 91°35′W, 119 m).
Topotype—CHUB IV G, 0–7 cm.

Diagnosis: Test small in size, 4–$5^1/_2$ subglobular chambers in the final whorl, about 10–13 chambers in all arranged in a very low trochospire of about $2^1/_2$ whorls, with no evidence of a peripheral keel. Chambers from the spiral side reniform, slightly inflated giving a lobulated periphery, sutures slightly curved, depressed; from the ventral side, rounded-subquadrate, well-inflated, sutures curved, depressed. Aperture interiomarginal, umbilical-extraumbilical, a low but fairly long arch overhung by a large plate-like lip on the base of the final chamber, and often by the lip of the penultimate chamber. Wall calcareous, finely but irregularly perforated, non-spinose.

Remarks: PARKER (1967) distinguished this species from *G. scitula* by its more lobulate periphery, rounded, more inflated chambers and more highly enclosed aperture.

Distribution: PARKER (1967) noted it from Late Pliocene (N. 21) to Recent sediments and plankton tows from the Gulf of California; it seems to prefer warm waters.

49. 1 *Globorotalia cavernula* BÉ, 1967
Plate 49, Figs. 1a–c, × 178; 1d, × 1428

Globorotalia cavernula BÉ, 1967, pp. 128–132, pl. 10, figs. 1–6. — BOLTOVSKOY, 1969, p. 252, pl. 2, figs. 13–14. — RÖGL and BOLLI, 1973, p. 568, pl. 7, figs. 5–6.

Types: Holotype—Eltanin 15-23-843, 250–500 m plankton tow (Lat. 55° 54′S, Long. 139° 56′E).

Plesiotype—RC8-38, trigger core top (Lat. 41° 53′S, Long. 37° 49′E, 3,784 m).

Diagnosis: Test small to medium in size, 5–6 rhomboid chambers in the final whorl, 13–18 chambers in all loosely arranged into a very low trochospire of about 3 whorls, with a thin peripheral keel. Chambers from the spiral side, reniform, well separated, slightly imbricated, very weakly inflated, sutures curved and slightly depressed; from the ventral side, subtriangular to nearly subquadrate, moderately inflated, sutures weakly depressed, radial. Aperture interiomarginal, umbilical-extraumbilical, a high, rounded arch with a thin rim-like opening onto the deep and wide umbilicus formed by the steep umbilical faces of the earlier chambers. Wall calcareous, thin, very finely perforated, slightly pustulate.

Remarks: This species is characterized by its very wide and open umbilicus, and, to a lesser extent, its imbricate chamber arrangement on the spiral side.

Distribution: Recent sediments and plankton tows from cool temperate waters of the Indo-Pacific (BÉ, 1967) and the Caribbean (RÖGL and BOLLI, 1973).

49. 2 *Globorotalia bermudezi* RÖGL and BOLLI, 1973
Plate 49, Figs. 2a–c, × 178; 2d, × 1428

Globorotalia (Turborotalia) sp. REISS, MERLING-REISS and MOSHKOVITZ, 1971, p. 157, pl. 10, figs. 1–2.
Globorotalia bermudezi RÖGL and BOLLI, 1973, pp. 567–568, pl. 6, figs. 16–20; pl. 16, figs. 1–3; text-figs. 6a–c.

Types: Holotype—DSDP Leg 15, Site 147, Core 2 (4.00 m subbottom depth), Section 1, top (Lat. 10° 42. 48′N, Long. 65° 10.48′W).

Plesiotype—V12-98, trigger core, 0–10 cm (Lat. 10° 47.1′N, Long. 65° 06.8′W, 736 m).

Diagnosis: Test small in size, $5^{1}/_{2}$–6 wedge-shaped chambers in the final whorl, 15 chambers in all arranged in a very low biconvex trochospire of about 3 whorls, with a thin imperforate periphery but no true keel. Chambers from the spiral side reniform, very weakly inflated, size increases moderately as added, sutures depressed and curved; from the ventral side they are subtriangular, moderately inflated, with radial, incised sutures. Aperture interiomarginal, umbilical-extraumbilical, a long, low arch with a wide flap-like lip, often partially attached to the lip of the previous chamber. Umbilicus wide and open. Wall calcareous, thin, very finely perforated, mostly towards the periphery, very pustulate on the ventral side.

Remarks: This species differs from *G. cavernula* by its less open umbilicus and more tightly packed chambers, and from *G. scitula* by a more open umbilicus and more radially expanded chamber.

Distribution: Reported from the Caribbean, Late Pleistocene to Recent by RÖGL and BOLLI (1973).

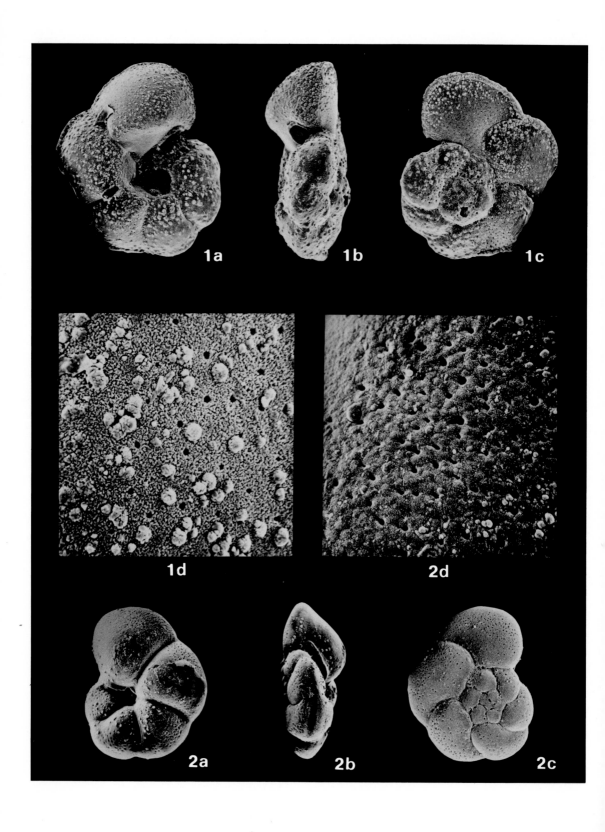

50. 1 *Globorotalia menardii* (Parker, Jones and Brady), 1865
Plate 50, Figs. 1a–c, × 53; 1d, × 714

Rotalia (Rotalie) menardii d'Orbigny, 1826, p. 273, no. 26 (*nomen nudum*, Parker, Jones and Brady, 1865).

Rotalia nitida d'Orbigny, 1826, p. 274, no. 31 (*nomen nudum*).

Rotalia nitida Fornasini, 1902, p. 66, pl. 3, fig. 4. — Banner and Blow, 1960, p. 33, pl. 6, fig. 3 (lectotype).

Rotalina (Rotalina) cultrata d'Orbigny, 1839a, p. 90, pl. 5, figs. 7–9. — Banner and Blow, 1960a, p. 34, pl. 6, fig. 1 (neotype).

Rotalina (Rotalina) menardii (d'Orbigny). — d'Orbigny, 1839c, p. 72 (synonymy only).

Rotalina cultrata d'Orbigny. — Bailey, 1851, p. 11, figs. 14–16.

Pulvinulina repanda (Fichtel and Moll) var. *menardii* Parker and Jones, 1865, pp. 391, 393, 394, pl. 16, figs. 35–37.

Pulvinulina repanda (Fichtel and Moll) var. *menardii* (d'Orbigny) subvar. *nitida* (Reuss). — Parker and Jones, 1865, p. 393 (listed).

Rotalia menardii Parker, Jones, and Brady, 1865, p. 20, pl. 3, fig. 81. — Banner and Blow, 1960, pls. 31–33, pl. 6, figs. 2a–c (lectotype).

Pulvinulina repanda (Fichtel and Moll) var. *menardii* (d'Orbigny) subvar. *cultrata* (d'Orbigny). — Parker and Jones, 1865, p. 393 (listed).

Discorbina sacharina Schwager, 1866, pp. 257–258, pl. 7, fig. 106.

Rotalina menardii (d'Orbigny). — Terquem, 1875, p. 27, pl. 3, figs. 1a–b.

Pulvinulina menardii (d'Orbigny) var. *cultrata* (d'Orbigny). — van den Broeck, 1876, p. 141, pl. 3, figs. 13,15.

Pulvinulina menardii (d'Orbigny). — Owen, 1868, p. 148, pl. 5, fig. 16. — Brady, 1884, p. 690, pl. 103, figs. 1,2.

Globorotalia menardii (d'Orbigny). — Cushman, 1927, p. 175. — Cushman, 1931, p. 91, pl. 17, fig. 1. — Bé, McIntyre and Breger, 1966, pp. 885–896, pls. 1–17.

Globorotalia (Pulvinulina) menardii (d'Orbigny). — Heron-Allen and Earland, 1932, p. 428.

Globorotalia cultrata (d'Orbigny). — Waller and Polski, 1959, p. 125, pl. 10, fig. 3.

Globorotalia (Globorotalia) cultrata menardii (Parker, Jones and Brady). — Blow, 1969, pp. 359–360, pl. 6, figs. 9–11.

Globorotalia (Menardella) menardii (d'Orbigny). — Bandy, 1972, pp. 297, 310.

Globoroialia menardii jamesbayensis McCulloch, 1979, p. 420, pl. 177, fig. 6.

Types: Holotype (*menardii*)—Model No. 10 of d'Orbigny, from Rimini on the Adriatic Sea.

Lectotype (*menardii*)—From Brady materials, off Laxey, Isle of Man, 15 fm.

Plesiotype (*menardii*)—V3-144B, trigger core top (Lat. 24°13′N, Long. 81° 52′W, 589 m).

Syntype (*cultrata*)—Recent marine sands of Cuba, Martinique, Guadaloupe, Jamaica.

Neotype (*cultrata*)—Earland collection, sample off Cape Cruz, Cuba.

Diagnosis: Test medium to quite large in size, 5–6 wedge-shaped chambers in the final whorl, about 15 chambers in all coiled in a biconvex trochospire of about $2^1/_2$ whorls, with a prominent peripheral keel. Chambers from the spiral side are semi-circular, nearly flat, with curved limbate sutures; from the umbilical side they are triangular, slightly inflated, not embracing, with depressed radial sutures. Aperture interiomarginal, umbilical-extraumbilical, a low arch with a large plate-like umbilical tooth. Wall calcareous, densely perforated with irregularly sized and shaped pores, non-spinose, typically possessing pustules near the umbilicus on the ventral side.

Remarks: BANNER and BLOW (1960) and PARKER (1962) have ably surveyed the controversy regarding the nomenclatural priority of *menardii* and *cultrata*. Various articles of the ICZN can be invoked to support either name. Here we are following the principle of common usage. *G. menardii* can be separated from *G. tumida* following PARKER (1962) by its larger diameter, thinner width and smoother shell surface. Observations by the authors in September 1974 on the lectotype of *R. menardii* selected by BANNER and BLOW (1960a) indicated that keel was more "fimbriated" and the apertural lip was larger than they figured, while the neotype of *R. cultrata* has a very smooth keel and greatly flattened but slightly inflated chambers. See also HEMLEBEN *et al.* (1977) for a detailed analysis of chamber formation.

Distribution: Early Miocene (N. 11) to Recent, equatorial to warm temperate waters.

50. 2 *Globorotalia tumida* (BRADY), 1877
Plate 50, Figs. 2a–c, × 53; 2d, × 714

Pulvinulina menardii (D'ORBIGNY) var. *tumida* BRADY, 1877, p. 535 (no figures) — BANNER and BLOW, 1960a, pp. 26–27, pl. 5, fig. 1 (lectotype).
Pulvinulina tumida BRADY. — BRADY, 1884, p. 692, pl. 103, figs. 4–6.
Globorotalia tumida (BRADY). — CUSHMAN, 1927, p. 91, pl. 19, fig. 12 (after BRADY). — WISEMAN and OVEY, 1950, p. 69, pl. 3, figs. 2a–c. — BOLLI, LOEBLICH and TAPPAN, 1957, pp. 41–42, pl. 10, figs. 2a–c.
Globorotalia (Pulvinulina) tumida (D'ORBIGNY) — HERON-ALLEN and EARLAND, 1932, p. 429.
Globorotalia menardii tumida (BRADY). — ERICSON and WOLLIN, 1956, pp. 117–118.
Globorotalia menardii (D'ORBIGNY) forma *tumida* — BOLTOVSKOY, 1968, p. 90, pl. 1, fig. 8.

Types: Syntype—From a soft, white calcareous rock which had been found by Liversidge (in Brady, 1877) on a beach on the east side of New Ireland (territory of New Guinea), Bismarck Archipelago.

Lectotype—From original materials

Plesiotype—Core V 28-241, 0–2 cm (Lat. 07°27′N, Long. 153°18′E, 4,488 m).

Diagnosis: Test medium to large in size, 6 wedge-shaped chambers in the final whorl, about 18 chambers in all arranged in a biconvex trochospire of about 3 whorls, with a distinct peripheral keel. Chambers from the spiral side are reniform, only slightly inflated, with raised, thickened, curving sutures; from the umbilical side they are sub-triangular, more inflated, with radially incised sutures. Aperture is interiomarginal, umbilical-extraumbilical, a low arch overhung by a large, plate-like lip. Wall calcareous, densely perforated with pores of varying sizes, non-spinose but coarsely pustulate on the ventral side near the umbilicus.

Remarks: Due to its close morphologic similarity to *G. menardii*, *G. tumida* has had a difficult time maintaining a separate identity. Their ancestral ties have continued to produce forms having characteristics often possessed by both. In the extreme, however, *G. tumida* is characteristically more elliptical and thicker than *G. menardii*, with a less lobulate periphery and a keel which is much wider at the start of the final whorl than at the end of it on the final chamber.

Distribution: Late Miocene (N. 18) to Recent, equatorial waters.

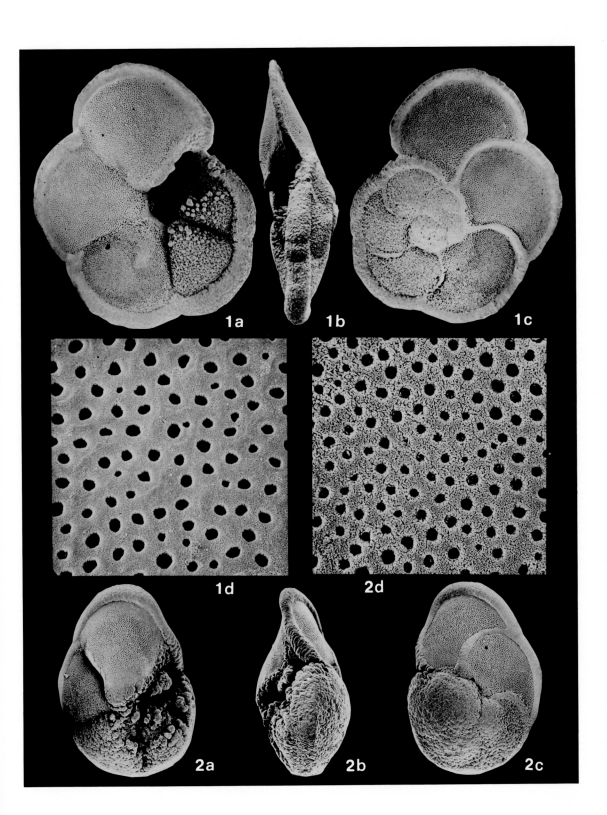

51. 1 *Globorotalia neoflexuosa* Srinivasan, Kennett and Bé, 1974
Plate 51, Figs. 1a–b, 2, × 40; 1c, × 666

Globorotalia menardii (d'Orbigny) linkage form to *G. menardii flexuosa* (Koch). — Boltovskoy, 1968, p. 90, pl. 1, fig. 7.
Globorotalia menardii flexuosa (Koch). — Bé and McIntyre (not Koch, 1923), 1970, pp. 595–601, figs. 1a–f. — Adegoke, Dessauvagie and Kogbe, 1971, p. 206, pl. 4, figs. 6–8.
Globorotalia menardii neoflexuosa Srinivasan, Kennett and Bé, 1974, pp. 321–324, pl. 1, figs. 1,2,8,9,10.
Globorotalia menardii (d'Orbigny) forma *neoflexuosa* Srinivasan, Kennett and Bé. — Adelseck, 1975, pp. 689–691, figs. 1d–n.
Globorotalia menardii (d'Orbigny) *gibberula* Bé, 1977, p. 61, pl. 12, figs. 36a–c.

Types: Holotype—Anton Brun 6-335-A-7201, 0–275 m plankton tow (Lat. 04°02′N, Long. 65°03′E).
Pleisiotype—RC12-328, 35 cm (Lat. 03°57′N, Long. 60°36′E, 3,087 m).

Diagnosis: Test large in size, 6–8 wedge-shaped chambers in the final whorl, about 12–15 chambers in all arranged into 3 whorls forming a biconvex trochospire with a distinct peripheral keel. Chambers from the spiral side are semicircular, flat or very weakly inflated, with curved sutures thickened by carina material; from the ventral side they are triangular, more inflated, with incised radial sutures. The final chamber, and often penultimate chamber, forms an acute angle relative to the plane of coiling of the earlier chambers. Aperture interiomarginal, umbilical-extraumbilical, a low arch shaded by a wide and heavy plate-like apertural lip formed at the base of the final chamber. Wall calcareous, densely perforated with pores of irregular shapes and sizes, non-spinose, coarsely pustulate near the umbilicus on the ventral side.

Remarks: Bé and McIntyre (1970), and later Srinivasan, Kennett and Bé (1974), have noted this species living in the Indian Ocean, the earlier paper by these authors suggesting that the form was conspecific with *G. tumida flexuosa*, whereas the later paper made the distinction that *G. t. flexuosa* was instead extinct. *G. neoflexuosa* differs from *G. menardii* by the acute angle of the final chamber and the earlier whorls, and from *G. flexuosa* by its larger diameter, less lobulate periphery and more uniform keel thickness. Adelseck (1975) suggested that *G. neoflexuosa* was an ecologic variant in upwelling areas of the eastern tropical Pacific. *G. menardii gibberula* Bé is interpreted to be an extreme flexuose form whose entire shell is arched.

Distribution: Recovered from plankton tows and Recent sediments of the Indian, Pacific and Atlantic Oceans, warm equatorial waters.

51. 2 *Globorotalia flexuosa* (Koch), 1923
Plate 51, Figs. 3a–c, × 40; 4, × 666

Pulvinulina tumida Brady var. *flexuosa* Koch, 1923, p. 351, text-figs. 9a–b, 10a–b.
Globorotalia menardii (d'Orbigny) var. *flexuosa* (Koch). — Boomgart, 1949, p. 145, pl. 10, fig. 9.
Globorotalia menardii flexuosa (Koch). — Ericson and Wollin, 1956, p. 117–118. — Ericson, Ewing, Wollin and Heezen, 1961, p. 203, pl. 3.
Globorotalia tumida (Brady) cf. *G. tumida* var. *flexuosa* (Koch). — Hamilton and Rex, 1959, p. 793, pl. 254, figs. 8–10.
Globorotalia tumida flexuosa (Koch). — Todd, 1964, p. 1094, pl. 294, fig. 4.
Globorotalia flexuosa (Koch). — Lamb and Beard, 1972, p. 52, pl. 12, figs. 6–8.

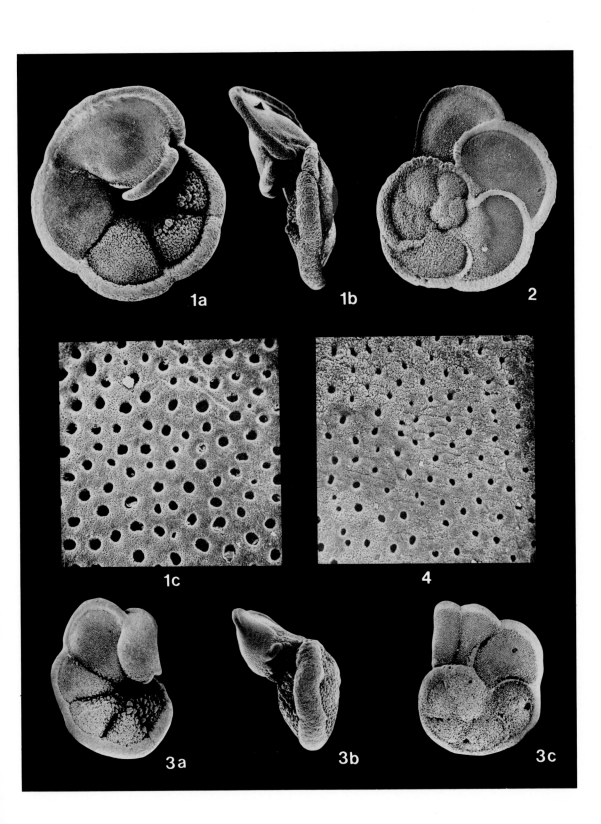

Types: Holotype—Along the road running north-south in Babad-Ngimbang-Kabu-Djombang, Northern Kabu, Surabaja, Java.

Plesiotype—V-26-127, 110 cm (Lat. 19°00′N, Long. 81°02′W, 6,251 m).

Diagnosis: Test large in size, about 6–7 wedge-shaped chambers in the final whorl, about 12 chambers in all arranged in a biconvex trochospire of about $2^{1}/_{2}$ whorls, with a thick peripheral keel. Chambers from the spiral side are reniform, slightly embracing, flat or very slightly inflated with slightly depressed, curving sutures, thickened with carina material; from the umbilical side they are subtriangular, moderately inflated, with incised radial sutures. The final, and often several earlier, chambers of the final whorl flex towards the umbilical side forming an angle relative to the plane of coiling of the earlier whorls. Aperture interiomarginal, umbilical-extraumbilical, a low arch usually overhung by a thick and wide plate-like lip formed at the base of the final chamber. Wall calcareous, coarsely perforated with pores of varying sizes and shapes, non-spinose, but coarsely pustulate near the umbilicus on the ventral side.

Remarks: This species has undergone much the same "identity crisis" as has *G. tumida*, having been considered everything from a bizarre form of *G. menardii* to a separate species, *G. flexuosa*. Works by Bé and McIntyre (1970) and later by Srinivasan *et al.* (1974) demonstrated that *G. flexuosa* was a species distinct from *G. neoflexuosa*, and that both of these were distinct from *G. menardii* and *G. tumida*. *G. flexuosa* differs from *G. tumida* by the flexing of the last few chambers toward the umbilical side of the test, and from *G. neoflexuosa* by its narrower, more elliptical outline, less lobulate periphery and peripheral keel which is wider on the exposed portions of the earlier whorls than on the final chamber.

Distribution: Bolli (1966) reported it as early as the *G. margaritae* zone (early Pliocene), and it has been observed to become extinct near the X-Y boundary of Ericson and Wollin (1968) in the Late Pleistocene. Probably equatorial.

52. 1 *Globorotalia pertenuis* BEARD, 1969
Plate 52, Figs. 1a–c, × 50; 1d, × 1000

Globorotalia pertenuis BEARD, 1969, pp. 552–553, pl. 1, figs. 1–6; pl. 2, figs. 5,6; pl. 3, fig. 4.

Types: Holotype—Texas A & M Univ., Sigsbee Knolls Core 64-A-9-5E, 190–192 cm (Lat. 23° 50′N, Long. 92° 24.5′W, 3,536 m).

Paratype—original sample, donated by J. H. BEARD.

Diagnosis: Test free, large, about 7 wedge-shaped chambers in the final whorl, 15–20 chambers in all arranged in a very low trochospire of about 3 whorls. Chambers partly embracing, sutures well incised, equatorial periphery lobulate. Test is carinate although the chambers of the final whorl tend to cover the carina of the earlier whorl on the spiral side. Aperture umbilical-extraumbilical, a low arch, greatly obscured by umbilical flaps at the base of each chamber coalescing in the umbilicus. Wall calcareous, finely perforated.

Remarks: This species is characterized by its very large size and numerous chambers, and is quite similar to *G. multicamerata* CUSHMAN and JARVIS from the Pliocene.

Distribution: Reported only from the Nebraskan interval of the Pleistocene, from the Gulf of Mexico.

52. 2 *Globorotalia lata* BRÖNNIMANN and RESIG, 1971
Plate 52, Figs. 2a–b, 3, × 50; 2c, × 1000

Globorotalia (Globorotalia) tumida (BRADY) *lata* BRÖNNIMANN and RESIG, 1971, pp. 1276–1277, pl. 29, fig. 3; pl. 49, figs. 4–5.

Types: Holotype—DSDP Hole 62-1, Core 4, Section 3, 15–17 cm (Lat. 01° 52.2′N, Long. 141° 56.3′E, 2,607 m).

Plesiotype—DSDP Hole 289, Core 4, core catcher (Lat. 00° 29. 92′S, Long. 158° 30.69′E, 2,206 m).

Diagnosis: Test medium in size, about 5–6 wedge-shaped chambers in the final whorl, about 12–15 chambers in all arranged in a biconvex trochospire of about 2–2$^1/_2$ whorls, with a heavy peripheral keel. Chambers from the spiral side, reniform, flat or very weakly inflated, with curved, slightly depressed sutures being slightly thickened with carina material; from the umbilical side, they are subtriangular, moderately inflated, with incised radial sutures; chamber size increases slowly. Aperture interiomarginal, umbilical-extraumbilical, a low circle overhung by a large plate-like lip formed at the base of the final chamber. Wall calcareous, densely perforated, non-spinose, but coarsely perforated near the umbilicus on the ventral side.

Remarks: This species combines features of both *G. menardii* and *G. tumida*, and, conceivably, may not represent a separate species but rather a deviant form of one or the other. It has been distinguished by BRÖNNIMANN and RESIG (1971) as having a "tangentially wider end chamber" than *G. tumida* and from *G. menardii* by its "tumid, heavy-keeled test."

Distribution: Reported only from the late Pliocene (N. 21) to the early Pleistocene (N. 22), in the western equatorial Pacific.

Plate 52 *Globorotalia pertenuis* BEARD, 1969
Globorotalia lata BRÖNNIMANN and RESIG, 1971

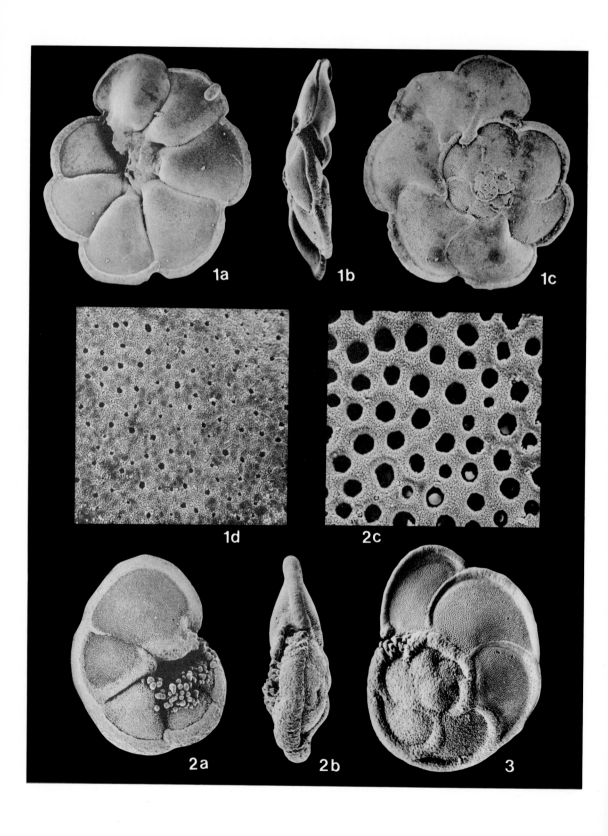

53. 1   *Globorotalia fimbriata* (Brady), 1884
Plate 53, Figs. 1a–c, × 66; 1d, × 1000

Pulvinulina menardii d'Orbigny var. *fimbriata* Brady, 1884, p. 691, pl. 103, fig. 3. — Banner and Blow, 1960a, p. 25, pl. 5, fig. 2 (lectotype).

Globorotalia menardii (d'Orbigny) var. *fimbriata* (Brady). — Cushman, 1931, p. 94, pl. 17, figs. 2a-b. — Wiseman and Ovey, 1950, p. 69, pl. 3, fig. 3.

Globorotalia fimbriata (Brady). — Hofker, 1956, p. 194, pl. 30, figs. 7–14. — Parker, 1967, p. 178, pl. 31, fig. 4.

Globorotalia menardii (d'Orbigny) forma *fimbriata* — Boltovskoy, 1968, p. 90.

Globorotalia akersi Snyder, 1975, pp. 302–304, pl. 1, figs. 1–6; pl. 2, figs. 1–5 [regarded herein as an aberrant morphotype of *G. fimbriata*].

Types: Syntypes—Bottom dredgings from 3 stations in the North Atlantic, 4 in the South Atlantic, 1 in the South Pacific.

Lectotype—Challenger Station 24 (Lat. 18°38′30″N, Long. 65°05′30″W, 390 fm).

Plesiotype—V15 Small Bio-trawl 52 (Lat. 18°45.5′N, Long. 66°30.5′W, 2,063–2,089 m).

Diagnosis: Test medium to large in size, 5 wedge-shaped chambers in the final whorl, about 10–12 chambers in all arranged into a biconvex trochospire of about 2 whorls, with a characteristic pustulate peripheral keel. Chambers from the spiral side are reniform to semicircular, flat or very weakly inflated, with curving depressed sutures slightly thickened by carina material; from the ventral side, they are subtriangular with depressed radial sutures. Chamber size increases slowly as added. Aperture interiomarginal, umbilical-extraumbilical, a very low, slit-like opening at the base of the umbilical shoulder, overhung by a large plate-like umbilical lip at the base of the final chamber, often extending to previous chambers. Wall calcareous, finely perforated, non-spinose, slightly pustulate near the umbilicus on the ventral side.

Remarks: This species is easily recognized by its pustulate (fimbriate) keel and shiny test surface, and is normally not easily confused with other members of the *menardii-tumida* group.

Distribution: Middle Pliocene (N. 20) to Recent, equatorial waters.

53. 2   *Globorotalia ungulata* Bermúdez, 1960
Plate 53, Figs. 2a, 3a–b, × 66; 2b, × 1000

Pulvinulina gilberti Bagg, 1908, p. 161, pl. 5, figs. 11–15.

Globorotalia cf. *G. menardii* (d'Orbigny). — Bradshaw, 1959, p. 44, pl. 8, figs. 10–12.

Globorotalia ungulata Bermúdez, 1961, p. 1304, pl. 15, figs. 6a-b. — Blow, 1969, p. 372, pl. 8, figs. 13–15. — Lamb and Beard, 1972, p. 57, pl. 11, figs. 7–9.

Globorotalia menardii ungulata Bermúdez. — Todd, 1964, p. 1093, pl. 295, fig. 3. — Adegoke, Dessauvagie and Kogbe, 1971, p. 206, pl. 4, figs. 14–17.

Types: Holotype—Atlantis Station 2953 (Lat. 21°47′N, Long. 84°32′30″W, 615 fm).

Near-Topotype—V17-20A, trigger core top (Lat. 24°02′N, Long. 91°25′W, 3,729 m).

Diagnosis: Test small to medium in size, about 5 wedge-shaped chambers in the final whorl, about 12–15 chambers in all arranged in a biconvex trochospire of $2^{1}/_{2}$ whorls, with a peripheral keel. Chambers from the spiral side, slightly reniform, flat or very neatly inflated, curved sutures slightly depressed but slightly limbate; from the ventral side, they are subtriangular, with incised radial sutures, distinctly inflated, the final chamber developing a high umbilical shoulder often with a thin ridge crest. Chambers increase slowly in size as added. Aperture interiomarginal, umbilical-extraumbilical, a very low slit-like opening at the base of the umbilical shoulder, overhung by a large plate-like umbilical lip. Wall calcareous, thin, densely perforated, non-spinose, moderately pustulate near the umbilicus on the ventral side.

Remarks: The high umbilical face and thin, shiny test wall make this species easy to identify, even with very small specimens.

Distribution: Reported from Late Pliocene (N.21) to Recent, equatorial waters.

54. 1 *Globorotalia truncatulinoides* (D'ORBIGNY), 1839
Plate 54, Figs. 1a–c, × 50; 1d, × 1000

Rotalina truncatulinoides D'ORBIGNY, 1839, p. 132, pl. 1, ? figs. 25–27.

Pulvinulina repanda (FICHTEL and MOLL) var. *menardii* (D'ORBIGNY) subvar. *truncatulinoides* (D'ORBIGNY).
— PARKER and JONES, 1865, pp. 393, 396, pl. 16, figs. 41–43 (not pl. 14, fig. 16).

Pulvinulina micheliniana (D'ORBIGNY) BRADY, 1884 (not D'ORBIGNY, 1840), p. 694, pl. 104, figs. 1a–c, 2a–b.

Pulvinulina truncatulinoides (D'ORBIGNY). — RHUMBLER, 1901, pp. 13,17–18, pl. 18, text-figs. 16a–b, 17, 18.

Globorotalia truncatulinoides (D'ORBIGNY). — CUSHMAN, 1927, p. 176. — CUSHMAN, 1931, p. 97, pl. 17, fig. 4. — BÉ and LOTT, 1964, pp. 823–824, figs. 1–3. — TAKAYANAGI, NIITSUMA and SAKAI, 1968, pp. 141–170, pl. 20–31. — GLAÇON, VERGNAUD-GRAZZINI and SIGAL, 1971, pp. 555–582, pls. 1–7.

Globorotalia (Pulvinulina) truncatulinoides (D'ORBIGNY). — HERON-ALLEN and EARLAND, 1932, p. 428.

Globorotalia (Globorotalia) truncatulinoides truncatulinoides (D'ORBIGNY). — BLOW, 1969, pp. 403–405, pl. 5, figs. 10–12; pl. 49, fig. 6 (neotype).

Globorotalia (Globorotalia) truncatulinoides pachytheca BLOW, 1969, pp. 370, 405–408, pl. 5, figs. 13–15; pl. 48, figs. 1–5.

Globorotalia (Truncorotalia) truncatulinoides (D'ORBIGNY). — CUSHMAN and BERMÚDEZ, 1949, p. 35, pl. 6, figs. 22–24. — FLEISHER, 1974a, p. 1025. — TODD, 1964, p. 1096, pl. 293, fig. 2.

Truncorotalia truncatulinoides (D'ORBIGNY) var. *nana* BERMÚDEZ, 1960, p. 75, pl. 1, figs. 48–50.

Types: Syntypes—Near the Canary Islands (Lat. 28°29′N, Long. 16°30′W).
Neotype—Challenger materials off Gomera, Canary Islands.
Near-topotype—V23-95, trigger core top (Lat. 30°24.2′N, Long. 18°23′W, 4,598 m).

Diagnosis: Test medium to large in size, 5 rhomboid chambers in the final whorl, about 15 chambers in all arranged in a plano-convex trochospire of about 3 whorls, with a distinct peripheral keel. Chambers, from the spiral side nearly polygonal with practically flat sides and straight sutures, flat to slightly concave; from the ventral side, they are subtriangular, strongly inflated and conical with straight, depressed sutures. Aperture interiomarginal, umbilical-extraumbilical, a very low arch with a thin rim-like lip at the base of the umbilical shoulder of the final chamber; the umbilicus is narrow but quite deep. Wall calcareous, finely perforated, non-spinose, densely pustulate on the ventral side, particularly in the umbilical depression.

Remarks: BLOW (1969) has discussed the nomenclatural problems prior to his proposal of the neotype for this species. It can be distinguished from *G. tosaensis* by the presence of a peripheral keel, more conical equatorial profile and more angular spiral suture. GLAÇON, VERGNAUD-GRAZZINI and SIGAL (1971, pl. 3, fig. 1c; pl. 4, fig. 1e) figured pustules with multiple terminations.

Distribution: Middle Pleistocene (N. 22) (near the Olduvai Magnetic Event) to Recent in temperate waters. ERICSON, WOLLIN and WOLLIN (1954) found that the Recent population of *G. truncatulinoides* in the North Atlantic is subdivided into oceanic provinces which are distinguished by dominance of either right- or left-coiling tests. These changes in the coiling ratios were used by ERICSON and WOLLIN (1956) to make precise correlation between Quaternary sequences in different parts of the oceans.

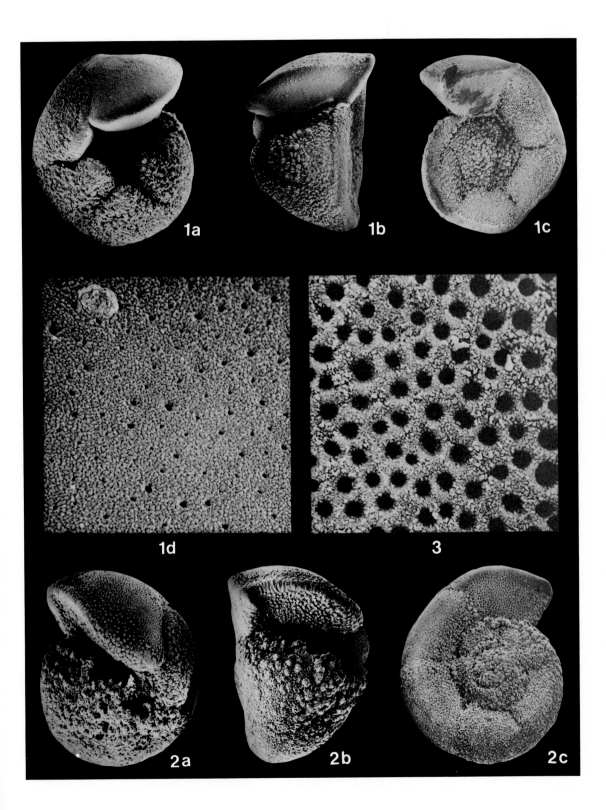

54. 2   *Globorotalia tosaensis* TAKAYANAGI and SAITO, 1962
Plate 54, Figs. 2a–c, × 83;   3, × 1000

Globorotalia tosaensis TAKAYANAGI and SAITO, 1962, pp. 81–82, pl. 28, figs. 11–12. — BLOW, 1969, pp. 393–394, pl. 4, figs. 10–12; pl. 40, figs. 4–7. — RÖGL, 1974, p. 748, fig. 3, nos. 10–1?; pl. 1, figs. 9–12.
Globorotalia (Turborotalia) tosaensis tenuitheca BLOW, 1969, pp. 357,394–396, pl. 4, figs. 13–17; pl. 40, figs. 1–3.
Globorotalia (Turborotalia) tosaensis tosaensis TAKAYANAGI and SAITO. — BLOW, 1969, pp. 357,393–394, pl. 4, figs. 10–12; pl. 40, figs. 4–7.
Globorotalia (Truncorotalia) tosaensis TAKAYANAGI and SAITO. — FLEISHER, 1974a, p. 1028.

Types: Holotype—Sample A-18, Nobori Formation, Nobori, Muroto City, Kochi Prefecture, Japan (Lat. 33°22′09″N, Long. 134°03′33″E).

Plesiotype—RC12-357, 1,034 cm (Lat. 08°58′N, Long. 120°14′E, 2,049 m).

Diagnosis: Test medium in size, 5 rhomboid chambers in the final whorl, about 10–12 chambers in all arranged in a low trochospire of about $2^{1}/_{2}$ whorls; no evidence of a keel is present. Chambers from the spiral side, subquadrate or polygonal, weakly inflated, very weakly lobulate, sutures nearly straight, spiral suture curved; from the ventral side, they are subtriangular, strongly inflated, sutures incised, slightly curved. Aperture interiomarginal, umbilical-extraumbilical, a moderately high, rounded arch with a thin rim-like lip. Wall calcareous, finely perforated, non-spinose, coarsely pustulate.

Remarks: This species can be distinguished from *G. truncatulinoides* primarily by its lack of a peripheral keel. This characteristic is particularly easy to discern on wet specimens, since specimens forming the transitional stages between *G. tosaensis* and *G. truncatulinoides* show a "buried" keel, on which it is not fully developed and moistening the specimen enhances this. Further differentiation is possible in that *G. tosaensis* has a curved spiral suture and a rounded equatorial periphery. It can also be differentiated from the *G. crassaformis* group by its rounder equatorial periphery.

Distribution: Late Pliocene (N. 4) to Late Pleistocene (N. 22) just above the Brunhes/Matuyama boundary, warm temperate waters. THOMPSON and SCIARRILLO (1978) placed the extinction level of this species at 590,000 years B.P.

55. 1   *Neoacarinina blowi* Thompson, 1973
Plate 55, Figs. 1a–c, × 120; 2, × 1200

Neoacarinina blowi Thompson, 1973, p. 470, pl. 1, figs. 1–5.
Globorotalia blowi (Thompson). — Poore and Berggren, 1975, p. 292, pl. 5, figs. 6–8.

Types: Holotype—RC8-39, 120 cm (Lat. 42° 53′S, Long. 42° 21′E, 4,330 m); holotype refigured.

Diagnosis: Test medium in size, 3–3$^1/_2$ globular chambers in the final whorl, up to 11 chambers in all arranged in a low trochospire of 3 whorls. Chambers subspherical to anguloconical, moderately inflated on the spiral side, strongly inflated on the ventral side, radially compressed, weakly lobulate, sutures radial, deeply incised. Aperture interiomarginal, umbilical-extraumbilical, a low arch overhung by a distinct rim-like lip. Wall calcareous, perforated, surface densely covered with multiple pustule groups.

Remarks: Since its proposal, this has been an unpopular species, due to its co-occurrence with and great similarity to *G. inflata* and also, perhaps more importantly, due to the concept of a genus and species defined by surface ultramicrostructure rather than the traditional chamber arrangement or mode of coiling. The basic morphology is very reminiscent of *G. inflata*, but, as was stated in Thompson (1973), co-occurring *G. inflata* do not possess multiple pustule groups (in the type description these pustule groups were termed spines for lack of a more concise phrase, but they by no means were then or are now intended to be collected in the genera assignment as *Globigerina* or *Hastigerina* type species).

Distribution: The type occurence is in the Crozet Basin of the Indian Ocean in a Late Pleistocene piston core, RC8-39. Since then we have observed it in other Indian Ocean core tops, several occurrences in the Pacific and two in the North Atlantic. More localities undoubtedly should be found, but misidentification of this species for *G. inflata* has hampered this.

55. 2   *Streptochilus tokelauae* (Boersma), 1969
Plate 55, Figs. 3a–c, × 330; 3d, × 3000

Bolivina tokelauae Boersma, 1969, p. 329, pl. 1, fig. 1.
Streptochilus tokelauae (Boersma). — Brönnimann and Resig, 1971, p. 1288, pl. 51, fig. 1. — Fleisher, 1974b, p. 993.

Types: Holotype—V18-262, 160 cm (Lat. 11° 57′S, 161° 26′W, 2,684 m).
    Plesiotype—DSDP Leg 30, Hole 289, Section 4, core catcher (Lat. 00° 29.92′S, Long. 158° 30.69′E, 2,206 m).

Diagnosis: Test very small in size, biserial, about 15 globular chambers in all. Chambers subspherical to quadrate, slowly increasing in size as added, partially embracing, sutures distinct, depressed. Aperture terminal, eccentric in position, a fairly high arch at the base of the final chamber, possessing, where preserved, a high collar with a thin rim-like lip. Wall calcareous, thin, shiny, very finely but irregularly perforated, nonspinose.

Remarks: Brönnimann and Resig (1971) raised the species identified by Boersma (1969) as a pelagic *Bolivina* to the type species of their new planktonic genus *Streptochilus*, expressing the feelings that 1) it was distinct from Paleogene *Chiloguembelina* and 2) its pelagic nature in Recent sediments had gone unstudied due to its very small size.

Distribution: Late Pliocene (N. 21) to Recent in equatorial Pacific sediments.

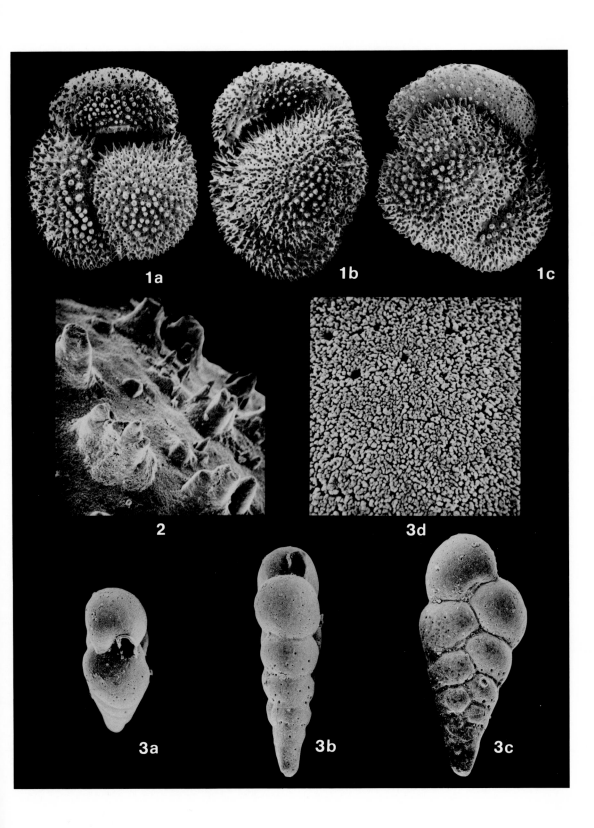

56   Aberrant Chamber Growth

Figs. 1   *Globorotalia menardii* (Parker, Jones and Brady), ×60.

Figs. 5   *Globorotalia inflata* (d'Orbigny), ×80.

Remarks: One type of aberrant growth encountered is that of grotesque chamber formation, where one or more, usually terminal, chambers distort, often making identification perplexing.

Fig. 2   *Pulleniatina obliquiloculata* (Parker and Jones), ×60.

Fig. 3   *Globorotalia scitula* (Brady), × 200.

Fig. 6   *Neogloboquadrina pachyderma* (Ehrenberg), ×120.

Remarks: Numerous doubled or "siamese twins" occur in pelagic sediments. Some of these have attained taxonomic status, such as "*Biorbulina bilobata*" in Plate 19, Fig. 5 or *G. ruber* "*helicina*" of this plate. The cause for such twinning is unknown, and can be considered to occur extremely rarely, although not infrequently in *Neogloboquadrina pachyderma* left-coiling variety.

Figs. 4   *Neogloboquadrina eggeri* (Rhumbler), ×120.

Figs. 7   *Globigerinoides ruber* (d'Orbigny) forma "*helicina,*" ×120.

Figs. 8   *Beella digitata* (Banner and Blow), ×80.

Remarks: Specimens are occasionally encountered possessing more chambers than "normal" populations of the species. Recognition of the taxa involved, however, should preclude any urge to erect new forms or varieties.

Plate 56   Aberrant Chamber Growth

GUIDE TO SYNONYMS

BIBLIOGRAPHY

ADEGOKE, O. S., DESSAUVAGIE, T. F. J. and KOGBE, C. A.
  1971   "Planktonic foraminifera in Gulf of Guinea sediments." Micropaleontology, vol. 17, no. 2, pp. 197–213, pls. 1–4.
ADELSECK, C. G., JR.
  1975   "Living *Globorotalia menardii* (D'ORBIGNY) forma *flexuosa* from the eastern tropical Pacific Ocean." Deep-Sea Res., vol. 22, pp. 689–691, fig. 1.
A. G. I. P. MINERARIA AUCT.
  1957   Foraminiferi Padani, Tavole 1–52.
AKERS, W. H.
  1955   "Some planktonic foraminifera of the American Gulf Coast and suggested correlations with the Caribbean Tertiary." Jour. Pal., vol. 29, no. 4, pp. 647–664, pl. 65, 3 text-figs.
  1972   "Planktonic foraminifera and biostratigraphy of some Neogene formations, northern Florida and Atlantic coastal plain." Tulane Stud. Geol. Paleont., vol. 9, nos. 1–4, pp. 1–140, pls. 1–60.
AKTÜRK, S. E.
  1976   "Traumatic variation in the *Globorotalia menardii* D'ORBIGNY group in late Quaternary sediments from the Caribbean." Jour. Foram. Res., vol. 6, no. 3, pls. 1–2.
ALLEE, W. C. and SCHMIDT, K. P.
  1951   "Ecological animal geography." New York: John Wiley & Sons, 2nd ed., pp. 1–715, text-figs. 1–139.
ANDERSON, O. R. and BÉ, A. W. H.
  1976a   "The ultrastructure of a planktonic foraminifer, *Globigerinoides sacculifer* (BRADY), and its symbiotic dinoflagellates." Jour. Foram. Res., vol. 6, no. 1, pp. 1–21, pls. 1–9.
  1976b   "A cytochemical fine structure study of phagotrophy in a planktonic foraminifer, *Hastigerina pelagica* (D'ORBIGNY)." Biol. Bull., vol. 151, pp. 437–449, text-figs. 1–10.
ANDROSOVA, V. P.
  1962   "Foraminifera from the bottom deposits of the western part of the Polar Basin." Moscow: Vses. Naucho. Issled. Inst. Morsk. Rybn. Khoz. Okeanogr. (VNIRD), Investigations for the International Geophysical Year Program., Sb. 1, Trudy, vol. 46, pp. 102–117, text-figs. 1–17, tables 1–4 [Russian].
ASANO, K.
  1957   "The foraminifera from the adjacent seas of Japan, collected by the S. S. Soyo-maru, 1922–1930; Part 3. Planktonic foraminifera." Tohoku Univ., Sci. Repts., 2nd ser. (Geol), vol. 28, pp. 1–26, pls. 1–2, text-fig. 1.
ASANO, K., INGLE, J. C., JR. and TAKAYANAGI, Y.
  1968   "Origin and development of *Globigerina quinqueloba* Natland in the North Pacific." Tohoku Univ., Sci. Repts., 2nd ser. (Geol), vol. 39, no. 3, pp. 213–241, 16 figs.
BAGG, R. M., JR.
  1908   "Foraminifera collected near the Hawaiian Islands by the steamer Albatross in 1902." U. S. Nat. Mus., Proc., vol. 34, no. 1603, pp. 113–172, pl. 5.
BAILEY, J. W.
  1851   "Microscopical examination of soundings, made by the U.S. Coast Survey off the Atlantic coast of the U.S." Smithsonian Contribution to Knowledge, vol. 2, art. 3, pp. 1–15.
BANDY, O. L.
  1960   "The geologic significance of coiling ratios in the foraminifer *Globigerina pachyderma* (EHRENBERG). Jour. Paleont., vol. 34, no. 4, pp. 671–681, text-figs. 1–7.
  1972a   "Environmental significance of variations in *Globigerina bulloides* D'ORBIGNY." 24th IGC., Abstract, p. 214.
  1972b   "Origin and development of *Globorotalia (Turborotalia) pachyderma* (EHRENBERG)." Micropaleontology, vol. 18, no. 3, pp. 294–318, pls. 1–8, text-figs. 1–3.
BANDY, O. L., FRERICHS, W. and VINCENT, E.
  1967   "Origin, development, and geologic significance of *Neogloboquadrina* BANDY, FRERICHS, and VINCENT, gen. nov." Cushman Found. Foram. Res., Contr., vol. 18, pt. 4, pp. 152–157, pl. 14, text-fig. 1.
BANDY, O. L., INGLE, J. C., JR. and FRERICHS, W. E.
  1967   "Isomorphism in *Sphaeroidinella* and *Sphaeroidinellopsis*." Micropaleontology, vol. 13, no. 4, pp. 483–488, pl. 1.
BANERJI, R. K., SCHAFER, C. T. and VINE, R.
  1971   "Environmental relationships and distribution of planktonic foraminifera in the equatorial and

northern Pacific waters." Atlantic Oceanogr. Lab., Bedford Inst., Report 1971–1977, 65 pp., 6pls.

BANNER, F. T. and BLOW, W. H.

1959   "The classification and stratigraphical distribution of the Globigerinaceae, Part I." Palaeontology, vol. 2, pt. 1, pp. 1–27, pls. 1–3.

1960a   "Some primary types of species belonging to the superfamily Globigerinaceae." Cushman Found. Foram. Res. Contr., vol. 11, pt. 1, pp. 1–41, pls. 1–8, text-figs. 1–2.

1960b   "The taxonomy, morphology and affinities of the genera included in the subfamily Hastigerininae." Micropaleontology, vol. 6, no. 1. pp. 19–31, text-figs. 1–11.

1965   "Progress in the planktonic foraminiferal biostratigraphy of the Neogene." Nature, vol. 208, no. 5016, pp. 1164–1166.

1967   "The origin, evolution and taxonomy of the foraminiferal genus *Pulleniatina* CUSHMAN, 1927." Micropaleontology, vol. 13, no. 2, pp. 133–162, pls. 1–4, 14 text-figs.

BÉ, A. W. H.

1959   "Ecology of Recent planktonic foraminifera; Part 1. Areal distribution in the western North Atlantic." Micropaleontology, vol. 5, no. 1, pp. 77–100, pl. 1–2, text-figs. 1–52, tables 1–2.

1960   "Some observations on Arctic planktonic foraminifera." Cushman Found. Foram. Res., Contr., vol. 11, pt. 2, pp. 64–68, 1 table, text-fig. 1.

1965   "The influence of depth on shell growth in *Globigerinoides sacculifer* (BRADY)." Micropaleontology, vol. 11, no. 1, pp. 81–97, pl. 1–2.

1967   "*Globorotalia cavernula*, a new species of planktonic foraminifera from the subantarctic Pacific Ocean." Cushman Found. Foram. Res., Contr., vol. 18, pt. 3, pp. 128–132, pls. 10–11.

1968   "Shell porosity of Recent planktonic foraminifera as a climatic index." Science, vol. 161, pp. 881–884, 3 figs., 1 table.

1969   "Microstructural evidence of the close affinity of *Globigerinella* CUSHMAN to *Hastigerina* THOMSON." In: BRÖNNIMANN, P. and RENZ., H. H. (eds.), Proceedings of the First International Conference on Planktonic Microfossils. Leiden: E. J. Brill, vol. 1, pp. 89–91, pls. 1–4.

1977   "An ecological, zoogeographic and taxonomic review of Recent planktonic foraminifera." In: RAMSAY, A. T. S. (ed.), Oceanic Micropaleontology. London: Academic Press, vol. 1, pp. 1–100.

BÉ, A.W.H. and ANDERSON, O. R.

1976   "Gametogenesis in planktonic foraminifera." Science, vol. 192, no. 4242, pp. 890–892.

BÉ, A. W. H. and ERICSON, D. B.

1963   "Aspects of calcification in planktonic foraminifera (Sarcodina)." Ann. N.Y. Acad. Sci., vol. 109, no. 1, pp. 65–81, figs. 1–10.

BÉ, A. W. H. and HAMLIN, W. H.

1967   "Ecology of Recent planktonic foraminifera; Part 3. distribution in the North Atlantic during the summer of 1962." Micropaleontology, vol. 13, no. 1, pp. 87–106, text-figs. 1–41, tables 1–3.

BÉ, A. W. H., HARRISON, S. M. and LOTT, L.

1973   "*Orbulina universa* D'ORBIGNY in the Indian Ocean." Micropaleontology, vol. 19, no. 2, pp. 150–192, pls. 1–10.

BÉ, A. W. H. and HEMLEBEN, C.

1970   "Calcification in a living planktonic foraminifera, *Globigerinoides sacculifer* (BRADY)". Neues Jahrb. Geol. Paläont., Abh., vol. 134, no. 3, pp. 221–234, text-fig. 1, pls. 25–32.

BÉ, A. W. H., HEMLEBEN, C., ANDERSON, O. R. and SPINDLER, M.

1980   "Pore structures in planktonic foraminifera." Jour. Foram. Res., vol. 10, no. 2, pp. 117–128, pls. 1–4.

BÉ, A. W. H., JONGEBLOED, W. L. and McINTYRE, A.

1969   "X-ray microscopy of Recent planktonic foraminifera," Jour. Pal., vol. 43, no. 6, pp. 1384–1396, pls. 161–167, text-fig. 1.

BÉ, A. W. H. and LOTT, L.

1964   "Shell growth and structure of planktonic foraminifera." Science, vol. 145, no. 3634, pp. 823–824, text-figs 1–3.

BÉ, A. W. H. and McINTYRE, A.

1970   "*Globorotalia menardii flexuosa* (KOCH): An extinct foraminiferal subspecies living in the northern Indian Ocean". Deep-Sea Res., vol. 17, pp. 595–601, 1 pl, 3 figs., 2 tables.

Bé, A. W. H., McIntyre, A. and Breger, D.
  1966   "Shell microstructure of a planktonic foraminifera, *Globorotalia menardii* (d'Orbigny)." Eclogae
          Geol. Helv., vol. 59, no. 2, pp. 885–896, pls. 1–17, 2 text-figs, 1 table.
Bé, A. W. H. and Tolderlund, D. S.
  1971   "Distribution and ecology of living planktonic foraminifera in surface waters of the Atlantic and
          Indian Oceans." In: Funnell, B. M. and Riedel, W. R. (eds.), Micropaleontology of Oceans,
          Cambridge Univ. Press, pp. 105–149, text-figs. 601–627.
Beard, J. H.
  1969   "Pleistocene paleotemperature record based on planktonic foraminifera, Gulf of Mexico." Gulf
          Coast Assoc. Geol. Socs., Trans., vol. 19, pp. 535–553, pls. 1–3, text-figs. 1–5.
Belford, D. J.
  1962   "Miocene and Pliocene planktonic foraminifera, Papua-New Guinea". Australia, Bur. Min. Res.
          Geol. Geophys., Bull., no. 62–1, pp. 1–51, pls. 1–8, text-figs. 1–3, 1 table.
Belyaeva, N. V.
  1964   "Distribution of planktonic foraminifera in the water and on the floor in Indian Ocean." Trudy
          Inst. Okean., Acad. Nauk S.S.S.R., vol. 68, pp. 12–83, pls. 1–3, figs. 1–27. (in Russian).
Berger, W. H.
  1969   "Kummerform foraminifera as clues to oceanic environments." Amer. Assoc. Petrol. Geol., Bull.,
          vol. 53, no. 3, p. 706.
  1971   "Planktonic foraminifera: sediment production in an oceanic front." Jour. Foram. Res., vol. 1,
          no. 3, pp. 95–118.
Berger, W. H. and Soutar, A.
  1967   "Planktonic foraminifera: Field experiment on production rate." Science, vol. 156, no. 3781,
          pp. 1495–1497.
Bermúdez, P. J.
  1960   "Foraminiferos planctonicos del Golfo de Venezuela." Mem. Soc. Ciencias Nat. La Salle, vol. 20,
          no. 55, pp. 58–76, pl. 1.
  1961   "Contribución al estudio de las Globigerinidea de la región Caribe-Antillana (Paleoceno-Re-
          ciente)." Dir. Geol., Bol. Geol., Publ. Espec. 3 (Congr. Geol. Venezolano, III, 1960, Mem.) vol.
          3, pp. 1119–1393, pls. 1–20.
Bermúdez, P. J. and Bolli, H. M.
  1969   "Consideraciones sobre los sedimentos del Miocene medio al Receinte de las costas central y oriental
          de Venezuela Tercera parte Los foraminiferos planctonicos." Venezuela: Dir. Geol., Bol. Geol.,
          vol. 10, no. 20, pp. 137–223, pls. 1–18, text-figs. 1–6.
Bermúdez, P. J. and Seiglie, G. A.
  1963   "Estudio sistemático de los foraminíferos del Golfo de Cariaco." Oriente, Univ., Inst. Oceanogr.,
          Bol., Cumana, Venezuela, vol. 2, no. 2, pp. 1–267, pls. 1–29.
Blackwelder, R. E.
  1967   "Taxonomy." New York: John Wiley & Sons, pp. 1–698.
Blair, D.
  1965   "The distribution of planktonic foraminifera in Deep-Sea cores from the southern Ocean, Antarc-
          tica." Florida State Univ., Dept. Geol., Contr., no. 10, 141 pp.
Blow, W. H.
  1956   "Origin and evolution of the foraminiferal genus *Orbulina* d'Orbigny." Micropaleontology, vol. 2,
          no. 1, pp. 57–70, text-figs. 1–4.
  1959   "Age, correlation biostratigraphy of the upper Tocuyo (San Lorenzo) and Pozón Formations,
          Eastern Falcón, Venezuela." Bull. Amer. Paleont., vol. 39, no. 178, pp. 59–251, pls. 6–19.
  1969   "Late Middle Eocene to Recent planktonic foraminiferal biostratigraphy." In: Brönnimann, P.
          and Renz, H. H., (eds.), Proceedings of the First International Conference on Planktonic Micro-
          fossils. Leiden: E. J. Brill, vol. 1, pp. 199–422, pls. 1–54, text-figs. 1–43.
  1979   "The Cainozoic Globigerinida." Leiden: E. J. Brill, vols. 1–3, pp. 1–1413, pls. 1–264.
Blow, W. H. and Barner, F. T.
  1962   "The mid-Tertiary (Upper Eocene to Aquitanian) Globigerinaceae." In: Eames, F. E., Banner, F. T.,
          Blow, W. H. and Clarke, W. J., Fundamentals of Mid-Tertiary Stratigraphic Correlation. Cam-
          bridge Univ. Press, pp. 61–163, figs. 1–19, pls. 8–17.

BOERSMA, A.
1969   *In* KIERSTEAD, C. H., LEIDY, R. R. D., FLEISHER, R. L. and BOERSMA, A. "Neogene zonation of tropical Pacific cores." In: BRÖNNIMANN, P. and RENZ, H. H. (eds.), Proceedings of the First International Conference on Planktonic Microfossils. Leiden: E. J. Brill, vol. 2, pp. 328–338, pl. 1.

BOLLI, H. M.
1957   "Planktonic foraminifera from the Oligocene–Miocene Cipero and Lengua Formations of Trinidad, B.W.I." U.S. Nat. Mus., Bull., no. 215, pp. 97–123, pls. 22–29.
1966   "The planktonic foraminifera in well Bodjonegoro-1 of Java." Eclogae Geol. Helv., vol. 59, no. 1, pp. 449–461, pl. 1.
1970   "The foraminifera of sites 23–31, Leg 4." In: BADER, R. G., GERARD, R. D., *et al.*, Initial Reports of the Deep Sea Drilling Project, vol. 4, pp. 577–643, figs. 1–22, pls. 1–90.

BOLLI, H. M., LOEBLICH, A. R., JR. and TAPPAN, H.
1957   "Planktonic foraminiferal families Hantkeninidae, Orbulinidae, Globorotaliidae and Globotruncanidae." U.S. Nat. Mus., Bull., no. 215, pp. 3–50, pls. 1–11.

BOLLI, H. M. and PREMOLI-SILVA, I.
1973   "Oligocene to Recent planktonic foraminifera and stratigraphy of the Leg 15 sites in the Caribbean Sea." In: EDGAR, N. T., SAUNDERS, J. B., *et al.*, Initial Reports of the Deep Sea Drilling Project, vol. 15, pp. 475–497, figs. 1–15.

BOLTOVSKOY, E.
1959   "Foraminiferal as biological indicators in the study of ocean currents." Micropaleontology, vol. 5, no. 4, pp. 473–481.
1966   "La zona de Convergencia Subtropical/Subantartica en el Oceano Atlantico (parte occidental)." Argentina, Serv. Hidr. Nav., no. H-640, pp. 1–69, pl. 1.
1968   "Living planktonic foraminifera of the eastern part of the tropical Atlantic." Rev. Micropal., vol. 11, no. 2, pp. 85–98, pls. 1–2.
1969   "Living planktonic foraminifera at the 90°E meridian from the Equator to the Antarctic." Micropaleontology, vol. 15, no. 2, pp. 237–255, pls. 1–3.
1971a   "Planktonic foraminiferal assemblages of the epipelagic zone and their thanatocoenoses." In: FUNNELL, B. M. and RIEDEL, W. R. (eds.), The Micropaleontology of Oceans. Cambridge Univ. Press, pp. 277–288, 1 fig.
1971b   "Ecology of the planktonic foraminifera living in the surface layer of the Drake Passage." Micropaleontology, vol. 17, no. 1, pp. 53–68, pl. 1.
1974   "*Globorotalia hirsuta eastropacia* n. subsp.—planktonic subspecies (Foraminiferida) from the tropical Pacific Ocean." Rev. Espan. Micropal., vol. 6, no. 1, pp. 127–133, pl. 1, figs. 1–14, 1 map.
1976   "New observations on the solution of planktonic foraminiferal tests and spines." In: TAKAYANAGI, Y. and SAITO, T. (eds.), Progress in Micropaleontology, New York: Amer. Mus. Nat. Hist., Micropaleontology Press Spec. Pub., pp. 17–19.

BOLTOVSKOY, E. and WATANABE, S.
1975   "First record of *Globigerinoides obliquus* BOLLI in Recent bottom sediments." Jour. Foram. Res., vol. 5, no. 1, pp. 40–41, text-figs. 1a–2b.

BOOMGART, L.
1949   "Smaller foraminifera from Bodjonegoro (Java)." Utrecht Univ., doct. dissertation. Smit & Dontje, Sappemeer (Holland), pp. 1–175, pls. 1–14.

BRADSHAW, J. S.
1959   "Ecology of living planktonic foraminifera in the north and equatorial Pacific Ocean." Cushman Found. Foram. Res., Contr., vol. 10, pt. 2, pp. 25–46, pls. 6–8, text-figs. 1–43.

BRADY, H. B.
1877   "Supplementary note on the foraminifera of the Chalk (?) of the New Britain Group." Geol. Mag., London, 1877, n. s., decade 2, vol. 4, no. 12, pp. 534–546.
1878   "On the reticularian and radiolarian Rhizopoda (Foraminifera and Polycystina) of the North-Polar Expedition of 1875–76." Ann. Mag. Nat. Hist., London, ser. 5, vol. 1, pp. 425–440, pl. 20–21.
1879   "Notes on some of the reticularian Rhizopoda of the Challenger Expedition II: Additions to the knowledge of porcellaneous and hyaline types." Quart. Jour. Micro. Sci., London, n. s., vol. 19, pp. 20–63, 261–299, pls. 3–5.

1881   "On some Arctic foraminifera from soundings obtained on the Austro-Hungarian North-Polar Expedition of 1872–1874." Ann. Mag. Nat. Hist., London, ser. 5, vol. 8, no. 48, pp. 393–418, pl. 21.

1882   "Report on the foraminifera." In: TIZARD and MURRAY, J., Exploration of the Faröe Channel, during the summer of 1880, in H. M. S. Knight Errant, with subsidiary reports." Roy. Soc. Edinburgh, Proc., vol. 11 (1880–1882), no. 111, pp. 708–717.

1884   "Report on the foraminifera dredged by H.M.S. Challenger during the years 1873–1876." Challenger Expedition 1873–1876, Rept., London, Zool., pt. 22, vol. 9, pp. 1–814, pls. 1–115 (in Atlas).

BRODSKY, A. L.
1929   "The fauna of the basins of the desert Kara-Kum." Asia Media Acta, Univ. Tashkent, ser. 12a (Geogr.), fasc. 5, pp. 1–43, pl. 1.

BROECK, E. VAN DEN
1876   "Étude sur les foraminifères de la Barbade (Antilles)." Soc. Belge Micr., Ann., Brussells, vol. 1 (1875–1876), pp. 55–152, pls. 2–3.

BRÖNNIMANN, P. and RESIG, J.
1971   "A Neogene globigerinacean biochronologic time scale of the southwest Pacific." In: WINTERER et al., Initial Reports of the Deep Sea Drilling Project, vol. 7, pt. 2, pp. 1235–1469, pls. 1–51.

BUCKLEY, H. A.
1973   "Globorotalia (Clavatorella) oveyi n. sp., première mention Récente d'un sous-genre de Foraminifère du Néogène." Rev. Micropal., vol. 16, no. 3, pp. 168–172, pl. 1.

BURT, B. J. and SCOTT, G. H.
1975   "Spinosity and coiling geometry in Pulleniatina (Foraminiferida)." Jour. Foram. Res., vol. 5, no. 3, pp. 166–175, pls. 1–3, text-figs. 1–2.

BÜTSCHLI, O.
1880   "Protozoa." In: BRONN, H. G., Klassen und Ordnungen des Their-Reichs, Leipzig und Heidelberg: C. F. Winter, vol. 1, pt. 1 (1880–1882), pp. 1–224, pls. 1–13.

CARPENTER, W. B.
1862   "Introduction to the study of the foraminifera." London: Ray Soc., pp. 1–319, pls. 1–22.

CHAPAMN, F.
1902   "On the foraminifera collected from the Funafuti Atoll from shallow and moderately deep water." Linn. Soc. London, Jour. Zool., vol. 28 (1900–1903), no. 184, pp. 379–417, pl. 35–56.

CHRISTIANSEN, B. O.
1965   "A bottom form of the planktonic foraminifer Globigerinoides rubra (D'ORBIGNY, 1839)." Pubbl. Stez. Zool. Napoli, vol. 34, no. 2, pp. 197–202, text-figs. 1–3.

CIFELLI, R.
1961   "Globigerina incompta, a new species of pelagic foraminifera from the North Atlantic." Cushman Found. Foram. Res., Contr., vol. 12, pt. 3, pp. 83–86, 1 table, pl. 4.

1965   "Planktonic foraminifera from the western North Atlantic." Smithsonian Misc. Coll., vol. 148, no. 4, pp. 1–36, pls. 1–9.

1973   "Observations on Globigerina pachyderm (EHRENBERG) and Globigerina incompta CIFELLI from the North Atlantic." Jour. Foram. Res., vol. 3, no. 4, pp. 157–166, pls. 1–4.

1975   "Views and observations on the taxonomy of certain Neogene planktonic foraminiferal species." In: TAKAYANAGI, Y. and SAITO, T. (eds.), Progress in Micropaleontology. New York: Amer. Mus. Nat. Hist., Micropaleontology Press, pp. 36–46, pls. 1–3.

CIFELLI, R. and SMITH, R. K.
1970   "Distribution of planktonic foraminifera in the vicinity of the North Atlantic current." Smithsonian Contr. Paleobiol., no. 4, pp.1–52, pls. 1–6.

CITA, M. B., PREMOLI-SILVA, I. and ROSSI, R.
1965   "Foraminiferi planctonici del Tortoniano-tipo." Riv. Ital. Paleont. Strat., vol. 71, no. 1, pp. 217–308, pls. 18–31.

COLALONGO, M. L. and SARTONI, S.
1967   "Globorotalia Nirsuta aemiliana nuova sottospecie cronologica del Pliocene in Italia." Giorn. Geol., ser. 2, vol. 34, fasc. 1, pp. 265–284, pls. 30–31.

COLOM, G.
1952   "Foraminiferos de las costas de Galicia (Campañas del "Xauen" en 1949 y 1950)." Bol. Inst. Español Oceanogr., Madrid, no. 51, pp. 3–59, pls. 1–8, text-figs. 1–5, map, table.

CONATO, V. and FOLLADOR. U.
1967   "*Globorotalia crotonensis* e *Globorotalia crassacrotonensis* nuove specie del Pliocene italiano."
          Boll. Soc. Geol. Ital., vol. 86, fasc. 3, pp. 555–563, text-figs. 1–6.

CORDEY, W. G.
1967   "The development of *Globigerinoides ruber* (D'ORBIGNY, 1839) from the Miocene to Recent." Palae-
          ontology, vol. 10, pt. 4, pp. 647–659, pl. 103.

CUSHMAN, J. A.
1914   "A monograph of the foraminifera of the North Pacific Ocean; Part. IV. Chilostomellidae,
          Globigerinidae, Nummulitidae." U.S. Nat. Mus., Bull., vol. 71, no. 4, pp. 1–46, pls. 1–19.
1915   "A monograph of the foraminifera of the North Pacific Ocean; Part. V. Rotaliidae, U.S. Nat.
          Mus., Bull., no. 71, pp. 1–83, pls. 1–31.
1917   "New species and varieties of foraminifera from the Philippines and adjacent waters." U.S. Nat.
          Mus., Proc., vol. 51, no. 2172, pp. 651–662.
1921   "Foraminifera of the Philippine and adjacent seas." U.S. Nat. Mus., Bull., no. 100, vol. 4, pp. 1–608,
          pls. 1–99.
1927a   "Recent foraminifera from off the west coast of North America." Scripps Inst. Oceanogr., Bull.,
          Tech. Ser., vol. 1, no. 10, pp. 119–188, pls. 1–6.
1927b   "An outline of a re-classification of the foraminifera." Cushman Lab. Foram. Res., Contr., vol. 3,
          pt. 1, pp. 1–105, pls. 1–21.
1931   "The foraminifera of the Atlantic Ocean; Part VIII. Rotaliidae, Amphistegenidae, Calcarinidae,
          Cymbaloporettidae, Globorotaliidae, Anomalinidae, Planorbulinidae, Rupertinidae, and Homo-
          tremidae." U.S. Nat. Mus., Bull., no. 104, pt. 8, pp. 1–179, pls. 1–26.
1941   "The species described as *Globigerina* by D'ORBIGNY in 1826." Cushman Lab. Foram. Res., Contr.,
          vol. 17, pt. 2, pp. 38–42, pls. 10–12.

CUSHMAN, J. A. and BERMÚDEZ, P. J.
1949   "Some Cuban species of *Globorotalia*." Cushman Lab. Foram. Res., Contr., vol. 25, pt. 2, pp. 26–45,
          pls. 5–8.

CUSHMAN, J. A. and HENBEST, L. G.
1940   "Geology and biology of North Atlantic deep-sea cores between Newfoundland and Ireland. Part
          2. Foraminifera." U.S. Geol. Survey, Prof. Paper 196-A, pp. 35–56, pls. 8–10, text-figs.

CUSHMAN, J. A. and JARVIS, P. W.
1930   "Miocene foraminifera from Buff Bay, Jamaica." Jour. Paleont., vol. 4, pp. 353–368, pls. 32–34.

CUSHMAN, J. A. and STAINFORTH, R. M.
1945   "The foraminifera of the Cipero Marl Formation of Trinidad, British West Indies." Cushman Lab.
          Foram. Res., Spec. Pub., no. 14, pp. 3–75, pls. 1–16.

CUSHMAN, J. A., STEWART, R. E. and STEWART, K. C.
1930   "Tertiary Foraminifera from Humboldt County, California." San Diego Soc. Nat. Hist., Trans.,
          vol. 6, no. 2, pp. 41–94, pls. 1–8.

CUSHMAN, J. A., TODD, R. and POST, R. J.
1954   "Recent foraminifera of the Marshall Islands. In: Bikini and nearby atolls; Part. 2. Oceanography
          (biologic)." U.S. Geol. Survey, Prof. Paper 260-H, pp. 319–384, pl. 82–93, text-figs. 116–118,
          tables 1–5.

DESHAYES, G. P.
1832   "Encyclopedie methodique; Histoire naturelle des vers." Paris: Mme. v. Agasse, vol. 2, pt. 2,
          pp. 1–594.

DOLLFUS, G. F.
1905   "Review of: MILLET, F. W., "Report on the Recent foraminifera of the Malay Archipelago." " Rev.
          Critique Palézzol., Paris, 1905, annee 9, pp. 222–223.

DROOGER, C. W.
1953   "Miocene and Pleistocene foraminifera from Oranjestad, Aruba (Netherlands Antilles)." Cushman
          Found. Foram. Res., Contr., vol. 4, no. 4, pp. 116–147, pls. 19–24.

DROOGER, C. W. and KAASSCHIETER, J. P. H.
1958   "Foraminifera of the Orinoco-Trinidad-Paria Shelf." In: Reports of the Orinoco Shelf Expedition,
          Volume IV. K. Nederl. Akad. Wetensch., Verh., Afd. Natuurk., ser. 1, vol. 22, no. 1, pp. 7–108,
          pls. 1–5

EARLAND, A.
1934   "Foraminifera; Part III. The Falklands sector of the Antarctic (excluding South Georgia)." Dis-
        covery Reports, Cambridge, vol. 10 (1935), pp. 1–208, pls. 1–10, text-figs. 1–2.
EGGER. J. G.
1857   "Die foraminiferen der Miocän-Schichten bei Ortenburg in Nieder-Bayern." Neues Jahrb. Min.
        Geogn. Geol. Petref.-Kunde, Stuttgart, pp. 266–311, pls. 5–15.
1893   "Foraminiferen aus Meeresgrundproben, gelothet von 1874 bis 1876 von S. M. Sch. Gazelle."
        K. Bayer, Akad. Wiss. München, Math.-Physik. Cl., Abh. Bd. 18 (1895), Abth. 2, pp. 193–458,
        pls. 1–21.
EHRENBERG, C. G.
1843   "Verbreitung und Einfluss des mikroskopischen Lebens in Süd-und Nord-Amerika." K. Akad. Wiss.
        Berlin, Physik. Abh., Berlin, (Jahrg. 1841), Theil 1, pp. 291–445, pls. 1–4.
1845   "Über das kleinste organische Leben an mehreren bisher nicht untersuchten Erdpunkten; Mikros-
        kopische Lebensformen von Portugal und Spanien, Sud-Afrika, Hinter-Indien, Japan und Kurdis-
        tan." K. Preuss. Akad. Wiss. Berlin, Ber., Berlin, pp. 357–381.
1854a   "Das organische Leben des Meeresgrundes; Weitere Mittheilung über die aus grossen Meerestiefen
        gehobenen Grund-Massen; Charakteristik der neuen mikroskopischen Organismen des tiefen
        atlantischen Oceans." K. Preuss. Akad. Wiss. Berlin, Ber., pp. 34–75, 191–196, 213–251.
1854b   "Mikrogeologie," Leipzig, L. Voss, pp. 1–374, pls. 1–40.
1858   "Kurze Characteristik der 9 neuen Genera und der 105 neuen Species des ägäischen Meeres und
        des Tiefgrundes des Mittel-Meeres." K. Preuss. Akad. Wiss. Berlin, Monatsber., pp. 10–40.
1861   "Elemente des tiefen Meeresgrundes in Mexikanischen Golfstrome bei Florida; Über die Tief-
        grund-Verhältnisse des Oceans am Eingange der Davisstrasse und bei Island." K. Preuss. Akad.
        Wiss. Berlin, Monatsber., Berlin, pp. 222–240, 275–315, map, chart.
1872   "Mikrogeologische Studien als Zusammenfassung seiner Beobachtungen des kleinsten Lebens der
        Meeres-Tiefgründe aller Zonen und dessen geologischen Einfluss." K. Preuss. Akad. Wiss. Berlin,
        Monatsber., Berlin, pp. 265–322.
1873   "Microgeologische Studien über das kleinste Leben der Meeres-Tiefgründe aller Zonen und dessen
        geologischen Einfluss." K. Preuss Akad. Wiss. Berlin, Abh., Jahre. 1872, pp. 131–397, pls. 1–12.
1874   "Das unsichtbar wirkende Leben der Nordpolarzone." In: KOLDEWEY, K, Zweite deutsche nord-
        polarfahrt in 1869 und 1870. Leipzig, F. A. Brockhaus, vol. 2 (zool.), pp. 437–467, pl. 1.
EL-NAGGAR, E. R.
1971   "On the classification, evolution and stratigraphical distribution of the Globigerinaceae." In:
        FARINACCI, A. (ed.), Proceedings of the Second Planktonic Conference. Rome: Edizioni Tech-
        noscienza, vol. 1, pp. 421–476, pls. 1–7.
ERICSON, D. B.
1959   "Coiling direction of *Globigerina pachyderma* as a climatic index." Science, vol. 130, no. 3369,
        pp. 219–220.
ERICSON, D. B., EWING, M., WOLLIN, G. and HEEZEN, B. C.
1961   "Atlantic Deep-Sea Sediment Cores." Geol. Soc. America, Bull., vol. 72, pp. 193–206, pls. 1–3,
        50 figs.
ERICSON, D. B. and WOLLIN, G.
1956   "Correlation of six cores from the equatorial Atlantic and the Caribbean." Deep-Sea Res., vol. 3,
        pp. 104–125, 11 figs., 4 tables.
1968   "Pleistocene climates and chronology in deep-sea sediments." Science, vol. 162, no. 3859,
        pp. 1227–1234.
ERICSON, D. B., WOLLIN, G. and WOLLIN, J.
1954   "Coiling direction of *Globorotalia truncatulinoides* in deep-sea cores." Deep-Sea Res., vol. 2,
        pp. 152–158.
FICHTEL, L. VON and MOLL, J. P. C. VON
1798   "Testacea microscopica, aliaque minuta exq generibus Argonauta et Nautilus, ad naturam delineata
        et descripta (Mikroshopische und andere Kleine Schalthiere aus den Geshlechtern Argonaute und
        Schifter, nach der Gezeichnet und beschrieben.)" Wien, Camesina (1803 reprint), pp. 1–124, pl. 1–24.
1857   "Über die Organische Lebens Gormen in unerwartet grossen Tiefen des mittelmeeres." K. Preuss.
        Akad. Berlin, Monatsber., Berlin, pp. 538–570.

FLEISHER, R. L.
   1974a   "Cenozoic planktonic foraminifera and biostratigraphy, Arabian Sea, Deep Sea Drilling Project, Leg 23A." In: WHITMARSH, R. B., WESER, O. E., ROSS, D. A. *et al.*, Initial Reports of the Deep Sea Drilling Project, vol. 23, pp. 1001–1072, pls. 1–21, tables 1–2, 1 fig.
   1974b   "Preliminary report on late Neogene Red Sea foraminifera, Deep Sea Drilling Project, Leg 23B." In: DAVIES, T. A, LUYENDYKE, B. P. *et al*, Initial Reports of the Deep Sea Drilling Project, vol. 26, pp. 985–1011, pls. 1–2, tables 1–2, figs. 1–2.

FLINT, J. M.
   1899   "Recent foraminifera. A descriptive catalog of specimens dredged by the U.S. Fish Commission steamer Albatross." U.S. Nat. Mus., Ann. Rept. 1897, pt. 1, pp. 249–349, pls. 1–80.

FORNASINI, C.
   1898   "Globigerine adriatiche." R. Accad. Sci. Ist. Bologna, Mem. Sci. Nat., Bologna, ser. 5, vol. 7 (1897–1899), pp. 575–586, pl. 1–4.
   1899   "Le Globigerine fossili d'Italia." Palaeontogr. Italica, Mem. Pal., vol. 4, pp. 203–216, text-figs. 1–5.
   1902   "Sinossi metodica, dei foraminiferi sin qui rin venuti nella sabbia del Lido di Rimini." R. Accad. Sci. Ist. Bologna, Mem. Sci. Nat., ser. 5, vol. 10 (1902–1904), pp. 1–68, pl. 1–63.
   1903   "Sopra alcune specie di *"Globigerina"* institute da D'ORBIGNY nel 1826." R. Accad. Sci. Ist. Bologna, Bologna, N. S., vol. 7 (1902–1903), pp. 139–142, pl. 1.

FRERICHS, W. E.
   1968   "Pleistocene–Recent boundary and Wisconsin glacial biostratigraphy in the northern Indian Ocean." Science, vol. 159, no. 3822, pp. 1456–1458, tables 1–2, fig. 1.
   1970   "Distribution and ecology of benthonic foraminifera in the sediments of the Andaman Sea" Cushman Found. Foram. Res., Contr., vol. 21, pt. 4, pp. 123–147, text-figs. 1–15, tables 1–7.

FRERICHS, W. E., HEIMAN, M. E., BORGMAN, L. E. and BÉ, A. W. H.
   1972   "Latitudinal variations in planktonic foraminiferal test porosity: Part 1. Opitcal studies." Jour. Foram. Res., vol. 2, no. 1, pp. 6–13, tables 1–3, text-fig. 1–9.

GALLOWAY, J. J.
   1933   "A manual of foraminifera." Bloomington, Indiana: Principia Press, pp. 1–483, pls. 1–42.

GALLOWAY, J. J. and WISSLER, S. G.
   1927   "Pleistocene foraminifera from the Lomita Quarry, Palos Verdes Hills, California." Jour. Pal., vol. 1, pp. 35–87, pls. 7–12, tables 1, 2.

GLAÇON, G., VERGNAUD GRAZZWI, C. and SIGAL, M. J.
   1971   "Premier résultats d'une série d'observations saisonnières des foraminifères du plancton Méditerraneen." In: FARINACCI, A. (ed.), Proceedings of the Second Planktonic Conference. Rome: Edizioni Technoscienza, vol. 1, pp. 555–582, pls. 1–7.

HAMILTON, E. L. and REX, R. W.
   1959   "Lower Eocene phosphatized *Globigerina* ooze from Sylvania Guyot." U.S. Geol. Survey, Prof. Paper 260-W, pp. 785–797, pls. 250–254.

HARTONO, H. M. S.
   1964   "Coiling direction of *Pulleniatina obliquiloculata trochospira* n. var. and *Globorotalia menardii.*" Geol. Survey Indonesia, Bull., vol. 1, no. 1, pp. 5–12, text-figs. 3–6.

HEMLEBEN, C.
   1969   "Zur Morphogenese planktonischer Foraminiferen." Zitteliana, vol. 1, pp. 91–133, pls. 6–18, text-figs. 1–4.
   1975   "Spine and pustule relationships in some Recent planktonic foraminifera." Micropaleontology, vol. 21, no. 3, pp. 334–341, pls. 1–2.

HEMLEBEN, C., BÉ, A. W. H., ANDERSON, O. R. and TUNTIVATE, S.
   1977   "Test morphology, organic layers and chamber formation of the planktonic foraminifer *Globorotalia menardii* (D'ORBIGNY)." Jour. Foram. Res., vol. 7, no. 1, pp. 1–25, pls. 1–12.

HEMLEBEN, C., BÉ, A. W. H., SPINDLER, M. and ANDERSON, O. R.
   1979   ""Dissolution" effects induced by shell resorption during gametogenesis in *Hastigerina pelagica* (D'ORBIGNY)." Jour. Foram. Res., vol. 9, no. 2, pp. 118–124, pls. 1–2.

HERB, R.
   1968   "Recent planktonic foraminifera from sediments of the Drake Passage, southern Ocean." Eclogae Geol. Helv., vol. 61, no. 2, pp. 467–480, pls. 1–3.

HERMAN, Y.
1969   "Arctic Ocean Quaternary microfauna and its relation to paleoclimatology." Palaeogeogr., Palaeo-
        climat., Palaeoecol., vol. 6, no. 4, pp. 251–276, pls. 1–2, text-figs. 1–5, tables 1–2.
1974   "Arctic ocean sediments microfauna, and the climatic record in Late Cenozoic time." In: HERMAN,
        Y. (ed.), Marine Geology and Oceanography of the Arctic Seas. New York: Springer-Verlag,
        pp. 283–348, pls. 1–20, text-figs. 1–22, tables 1–7, maps.
1980a   "Globigerina exumbilicata HERMAN, 1974, a synonym of G. quinqueloba egelida CIFELLI and SMITH,
        1970." Jour. Pal., vol. 54, no. 3, p. 631.
1980b   "Globigerina cryophila, new name for Globigerina occlusa, HERMAN 1974." Ibid., p. 631.
HERON-ALLEN, E. and EARLAND, A.
1922   "Protozoa; Part II. Foraminifera." British Antarctic (Terra Nova) Exped. 1910, Nat. Hist. Rept.,
        London, Zool., vol. 6, no. 2, pp. 25–268, pls. 1–8, text-fig. 1.
1929   "Some new foraminifera from the South Atlantic." Roy. Micr. Soc. London, Jour., ser. 3, vol. 49,
        pt. 4, art. 27, pp. 324–334, pls. 1–4.
HOFKER, J.
1956   "Foraminifera Dentata, foraminifera of Santa Cruz and Thatch Island, Virgin Archipelago, West
        Indies." Spolia Zool. Mus. Kobenhavn., vol. 15, pp. 1–237, pls. 1–35.
1959   "On the splitting of Globigerina." Cushman Found. Foram. Res., Contr., vol. 10, pt. 1. pp. 1–9.
IACCARINO, S. and SALVATORINI, G.
1979   "Planktonic foraminiferal biostratigraphy of Neogene and Quaternary of Site 398 of DSDP Leg
        47B." In: VON RAD, U., RYAN, W. B. F. et al., Initial Reports of Deep Sea Drilling Project, vol. 47,
        pt. 2, pp. 255–286, pls. 1–9.
JENKINS, D. G.
1971   "New Zealand Cenozoic planktonic foraminifera." N. Z. Geol. Survey, Paleont. Bull., 42, pp. 1–
        278. pls. 1–23.
JENKINS, D. G. and ORR, W. N.
1972   "Planktonic foraminiferal biostratigraphy of the eastern Equatorial Pacific. DSDP Leg 9." In: HAYS,
        J. D. et al., Initial Reports of the Deep Sea Drilling Project, vol. 9, pp. 1059–1193, pls. 1–41, text-
        figs. 1–9, tables 1–2.
JONES, T. R. and PARKER, W. R.
1860   "On the rhizopodal fauna of the Mediterranean, compared with that of the Italian and some other
        Tertiary deposits." Geol. Soc. London, Quart. Jour., vol. 16, pp. 292–307.
1872   "On the foraminifera of the family Rotalinae (Carpenter) found in the Cretaceous formations; with
        notes on their Tertiary and Recent representatives." Geol. Soc. London, Quart. Jour., vol. 28, pt. 2,
        no. 110, pp. 103–131.
KANE, J.
1953   "Temperature correlations of planktonic foraminifera from the North Atlantic Ocean." Micro-
        paleontologist, vol. 7, no. 3, pp. 25–50, pls. 1–3.
KEANY, J. and KENNETT, J. P.
1972   "Pliocene–early Pleistocene paleoclimatic history recorded in Antarctic-Subantarctic deep-sea
        cores." Deep-Sea Res., vol. 19, no. 8, pp. 529–548, text-figs. 1–9.
KELLER, G.
1978   "Late Neogene biostratigraphy and paleo-oceanography of DSDP Site 310 Central North Pacific
        and correlation with Southwest Pacific." Marine Micropal., vol. 3, pp. 97–119, pls. 1–5.
KENNETT J.P.
1968a   "Globorotalia truncatulinoides as a paleo-oceanographic index." Science, vol. 159, no. 3822,
        pp. 1461–1463.
1968b   "Latitudinal variation in Globigerina pachyderma (EHRENBERG) in surface sediments of the southwest
        Pacific Ocean." Micropaleontology, vol. 14, no. 3, pp. 305–318, pl. 1.
1970   "Comparison of Globigerina pachyderma (EHRENBERG) in Arctic and Antarctic areas." Cushman
        Found. Foram. Res., Contr., vol. 21, pt. 2, pp. 47–49, pls. 8, 9.
1973   Middle and late Cenozoic planktonic foramiferal biostratigraphy of the southwest Pacific: DSDP
        Leg 21." In: BURNS R. E., ANDREWS, J. E. et al., Initial Reports of Deep Sea Drilling Project,
        vol. 21, pp. 575–640, pls. 1–21.
1976   "Phenotypic variation in some Recent and Late Cenozoic planktonic foraminifera." In: HEDLEY,

R. H. and ADAMS, C. G. (eds.), Foraminifera. London, Academic Press, vol. 2, pp. 111–170.

KOCH, R.
1923 "Die jungtertiäre Foraminiferenfauna von Kabu (Res. Surabaja, Java)." Eclogae Geol. Helv., vol. 18 (1923–1924), no. 2, pp. 342–361, pls. 1–11.

LAMB, J. L. and BEARD, J. H.
1972 "Late Neogene planktonic foraminifera in the Caribbean, Gulf of Mexico, and Italian stratotypes." Univ. Kansas, Paleont. Contr., no. 57 (Protozoa, art 8), pp. 1–67, pls. 1–36, text-figs. 1–25.

LE CALVEZ, Y.
1974 Révision des foraminifères de la collection D'ORBIGNY. I. Foraminifères des Iles Canaries." Cahiers de Micropal., no. 2, pp. 1–107, pls. 1–28.

LEE, J. J., FREUDENTHAL, H. D., KOSSOY, V. and BÉ, A. W. H.
1965 "Cytological observations on two planktonic foraminifera; *Globigerina bulloides* D'ORBIGNY, 1826 and *Globigerinoides ruber* (D'ORBIGNY, 1839) CUSHMAN, 1927." Jour. Protozool., vol. 12, no. 4, pp. 531–542, pls. 1–5.

LEROY, L. W.
1941 "Small foraminifera from the late Tertiary of the Nederlands East Indies." Colorado School Mines, Quart., vol. 36, no. 1, pt. 1, pp. 11–62, pls. 1–3; pt. 2, pp. 63–105, pls. 1–7; pt. 3, pp. 107–132, pls. 1–3.

LIDZ, L.
1966 "Deep-Sea Pleistocene biostratigraphy." Science, vol. 154, no. 3755, pp. 1448–1452, text-figs. 1–4.

LIPPS, J. H.
1966 "Wall structure, systematics, and phylogeny of planktonic foraminifera." Jour. Pal., vol. 40, no. 6, pp. 1257–1274, pl. 155, text-figs. 1–5, tables 1–2.

LOEBLICH, A. R., JR. and TAPPAN, H.
1957 "The new planktonic foraminiferal genus *Tinophodella*, and an emendation of *Globigerinita* BRÖNNIMANN." Washington Acad. Sci., Jour., vol. 47, no. 4, pp. 112–116, text-fig. 1.
1964 "Sarcodina, chiefly "thecamoebians" and Foraminiferida." In: MOORE, R. C. (ed.), Treatise on Invertebrate Paleontology. New York: Geol. Soc. Amer., pt. C, Protista 2, vols. 1–2, pp. C1–C900, text-figs. 1–653.

LOMNICKI, J. R. VON
1900 "Przyczyek do znajomósci fauny otwornic Miocenu Wieliczhi [Contribution to the knowledge of the foraminiferal fauna of the Miocene of Wieliczka]." Kosmos, LVOV, vol. 24 (1899), pp. 220–228, pl. 1 [Polish].
1901 "Einige Bemerkungen zum Aufsatze: Die Miocänen Foraminiferen in der Umgebung von Kolomea." Naturf. Ver. Brünn, Verh., Bd. 39 (1900), pp. 15–18.

McCULLOCH, I.
1979 "Qualitative observations on Recent foraminiferal tests with emphasis on the eastern Pacific" Univ. South. Calif., Allan Hancock Found., pt. 1, pp. 1–330; pt. 2, pp. 331–676; pt. 3, pp. 677–1078, pls. 47–248.

MAIYA, S., SAITO, T. and SATO, T.
1976 "Late Cenozoic planktonic foraminiferal biostratigraphy of northwest Pacific sedimentary sequences." In: TAKAYANAGI, Y. and SAITO, T. (eds.), Progress in Micropaleontology. New York: American Mus. Nat. Hist., Micropaleontology Press, pp. 395–422, pls. 1–6, text-figs. 1–7, tables 1–2.

MALMGREN, B. and KENNETT, J.
1972 "Biometric anlysis of phenotypic variation: *Globigerina pachyderma* (EHRENBERG) in the South Pacific Ocean." Micropaleontology, vol. 18, no. 2, pp. 241–248.
1976 "Biometric analysis of phenotypic variation in Recent *Globigerina bulloides* D'ORBIGNY in the southern Indian Ocean." Marine Micropal., vol. 1, no. 1, pp. 3–25, pl. 1.
1977 "Biometric differentiation between Recent *Globigerina bulloides* and *Globigerina falconensis* in the southern Indian Ocean." Jour. Foram. Res., vol. 7, no. 2, pp. 130–148, pl. 1.

MISTRETTA, F.
1962 "Foraminiferi planctonici del Pliocene inferiore di Altavilla Milicia (Palermo, Sicilia)" Riv. Ital. Pal. Strat., vol. 68, no. 1, pp. 97–114, pls. 8–11, text-fig. 1.

NATLAND, M. L.
1933 "The temperature- and depth-distribution of some Recent and fossil foraminifera in the Southern

California region." Univ. Calif, Scripps Inst. Oceanogr., Bull., Tech. Ser., vol. 3, no. 10, pp. 225–230.

1938   "New species of foraminifera from off the west coast of North America and from the later Tertiary of the Los Angels Basin." *Ibid.*, vol. 4, no. 5, pp. 137–164, pls. 3–7.

NATORI, H.

1976   "Planktonic foraminiferal biostratigraphy and datum planes in the Late Cenozoic sedimentary sequences in Okinawa-Jima, Japan." In: TAKAYANAGI, Y. and SAITO, T. (eds.), Progress in Micropaleontology. New York, Amer. Mus. Nat. Hist., Micropaleontology Press, Spec. Pub., pp. 214–243, pls. 1–5, text-figs. 1–3, tables 1–3.

OLSSON, R. K.

1974   "Pleistocene paleooceanography and *Globigerina pachyderma* (EHRENBERG) in Site 36, DSDP, Northeastern Pacific." Jour. Foram. Res., vol. 4, no. 2, pp. 47–60, pls. 1–3.

1976   "Wall structure, topography, and crust of *Globigerinapachyderma* (EHRENBERG)." In: TAKAYANAGI, Y. and SAITO, T. (eds.), Progress in Micropaleontology. New York: Amer. Mus. Nat. Hist., Micropaleontology Press, Spec. Pub., pp. 244–257, pls. 1–6.

ORBIGNY, A. D. D'

1826   "Tableau méthodique de la classe des Céphalopodes." Ann. Sci. Nat., Paris, ser. 1, vol. 7, pp. 96–314, pls. 10–17 (in Atlas).

1839a   "Foraminifères." In: SAGRA, R. DE LA, Histoire physique, politique et naturelle de l'Ile de Cuba. Paris: A. Bertrand, pp. 1–224, pls. 1–12 (plates published separately).

1839b   "Foraminifères des Iles Canaries." In: BARKER-WEBB, P. and BERTHELOT, S., Hist. Naturelle des Iles Canaries. Paris: Bethune, vol. 2, pt. 2, Zool., pp. 119–146, pls. 1–3.

1839c   "Voyage dans l'Amérique Méridionale—Foraminifères" Paris: Pitois-Leurault et Cᵉ; Strasbourg, V. Lebrault, vol. 5, pt. 5, pp. 1–86, pls. 1–9.

1846   "Foraminifères fossiles du bassin tertiaire de Vienne (Autriche) (Die fossilen foraminiferen des tertiären Beckens von Wien)." Paris, Gide et Comp., pp. 1–303, tables. 1–21.

ORR, W. W.

1969   "Variation and distribution of *Globigerinoides ruber* in the Gulf of Mexico." Micropaleontology, vol. 15, no. 3, pp. 373–379, pl. 1, text-figs. 1–6, table.

ORR, W. W. and ZAITZEFF, J. B.

1971   "A new planktonic foraminiferal species from the California Pliocene." Jour. Foram. Res., vol. 1, no. 1. pp. 17–19, pl. 1.

OWEN, S. R. I.

1868   "On the surface-fauna of mid-ocean; No. 2. Foraminifera." Linn. Soc. London, Jour., Zool., vol. 9 (1867), pp. 147–157, pl. 5.

PARKER, F. L.

1948   "Foraminifera of the continental shelf from the Gulf of Maine to Maryland." Harvard Coll., Mus. Comp. Zool., Bull., vol. 100, no. 2, pp. 213–241, pls. 1–7.

1954   "Distribution of the foraminifera in the northeastern Gulf of Mexico." *Ibid.*, vol. 111, no. 10, pp. 453–588, pls. 1–13.

1958   "Eastern Mediterranean foraminifera" Repts. Swedish Deep-Sea Exped., vol. VIII, Sediment Cores from the Meditteranean Sea and the Red Sea, no. 4, pp. 219–285, pls. 1–6, tables 1–18.

1960   "Living planktonic foraminifera from the Equatorial and Southeast Pacific" Tohoku Univ., Sci. Repts., 2nd Ser. (Geol), Spec. Vol., no. 4, pp. 71–82, text-figs. 1–20.

1962   "Planktonic foraminiferal species in Pacific sediments." Micropaleontology, vol. 8, no. 2, pp. 219–254, pls. 1–10.

1964   "Foraminifera from the experimental Mohole Drilling near Guadalupe Island, Mexico." Jour. Paleont., vol. 38, no. 4, pp. 617–636, pls. 97–102.

1965a   "Irregular distributions of planktonic foraminifera and stratigraphic correlation." In: SEARS, M. (ed.), Progress in Oceanography, vol. 3, pp. 267–272.

1965b   "A new planktonic species (Foraminiferida) from the Pliocene of Pacific deep-sea cores." Cushman Found. Foram. Res., Contr., vol. 16, pt. 4, pp. 151–152.

1967   "Late Tertiary biostratigraphy (planktonic foraminifera) of tropical Indo-Pacific deep-sea cores." Bull. Amer. Pal., vol. 52, no. 235, pp. 115–208, pls. 17–32, text-figs. 1–5, tables 1–4.

1976   "Taxonomic notes on some planktonic foraminifera." In: TAKAYANAGI, Y. and SAITO, T. (eds.), Progress in micropaleontology. New York, Amer. Mus. Nat. Hist., Micropaleontology Press, Spec. Pub., pp. 258–262, pl. 1.

PARKER, F. L. and BERGER, W. H.
1971   "Faunal and solution patterns of planktonic foraminifera in surface sediments of the South Pacific." Deep-Sea Res., vol. 18, pp. 73–107.

PARKER, W. K. and JONES, T. R.
1863   "On the nomenclature of the foraminifera; Part. X. The species enumerated by D'ORBIGNY in the "Annales des Sciences Naturelles," vol. 7, 1826." Ann. Mag. Nat. Hist., London, ser. 3, vol. 12, pp. 429–441.
1865   "On some foraminifera from the North Atlantic and Arctic Oceans, including Davis Straits and Baffin's Bay." Roy. Soc. London, Philos. Trans., London, vol. 155, pp. 325–441, pls. 13–19.

PARKER, W. K., JONES, T. R. and BRADY, H. B.
1865   "On the nomenclature of the foraminifera; Part. XII. The species enumerated by D'ORBIGNY in the "Annales des Sciences Naturelles," vol. 7, 1826." Ann. Mag. Nat. Hist., London, ser. 3, vol. 16, pp. 15–41, pls. 1–3.

PEZZANI, F.
1963   "Studio micropaleontologico di un campione della serie Messiniana di Tabiano Bagni (Parma)." Riv. Ital. Pal. Strat., vol. 69, no. 4, pp. 559–662, pls. 29–38.

PHLEGER, F. B. and PARKER, F. L.
1951   "Ecology of foraminifera, Northwest Gulf of Mexico; Part II. Foraminifera species." Geol. Soc. America, Mem. 46, pp. 1–64, pls. 1–20

PHLEGER, F. B., PARKER, F. L. and PEIRSON, J. F.
1953   "North Atlantic foraminifera." Repts. Swedish Deep-Sea Exped., Vol VII, Sediment Cores from the North Atlantic Ocean, no. 1, pp. 3–122, pls. 1–12.

POAG, C. W.
1972   "Neogene planktonic foraminiferal biostratigraphy of the western North Atlantic: DSDP Leg 11." In: HOLLISTER, C. D., EWING, J. I. et al., Intial Reports of Deep Sea Drilling Project, vol. 11, pp. 483–543, pls. 1–11.

POAG, C. W. and VALENTINE, P. C.
1976   "Biostratigraphy and ecostratigraphy of the Pleistocene basin Texas-Louisiana continental shelf." Gulf Coast Assoc. Geol. Socs., Trans., vol. 26, pp. 185–256, pls. 1–23.

POORE, R. Z.
1979   "Oligocene through Quaternary planktonic foraminiferal biostratigraphy of the North Atlantic: DSDP Leg 49." In: LUYENDYK, B. P., CANN, J. R. et al., Initial Reports of Deep Sea Drilling Project, vol. 49, pp. 447–518, pls. 1–20.

POORE, R. Z. and BERGGREN, W. A.
1975   "Late Cenozoic planktonic foraminiferal biostratigraphy and paleoclimatology of Hatton-Rockall Basin: DSDP Site 116." Jour. Foram. Res., vol. 5, no. 4, pp. 27–293, pls. 1–5, text-figs. 1–5.

REEVE, L.
1842   "Conchologia Systematica." London: Congman, Brown, Green and Congmens, vol. 2, pp. 1–337, pls. 130–300.

REISS, Z., MERLING-REISS, P., and MOSHKOVITZ, S.
1971   "Quaternary planktonic Foraminiferida and nannoplankton from the Mediterranean continental shelf and slope of Israel." Israel Jour. Earth-Sci., vol. 20, pp. 141–177, pls. 1–15.

REUSS, A. E.
1850   "Neue Foraminiferen aus den Schicten des Österreichischen Tertiärbeckens." K. Akad. Wiss. Wien, Math.-Nat. Cl., Denkschr., vol. 1, pp. 365–390, pls. 46–51.

RHUMBLER, L.
1895   "Entwurf eines natürlichen Systems der Thalamophoren." K. Ges. Wiss. Göttingen, Math.-Phys. Kl., Nachr., no. 1, pp. 51–98.
1901   "Nordische Plankton-Foraminiferen." In: BRANDT, K., (ed.), Nordisches Plankton. Kiel, Lipsius und Tischer, pt. 1 no. 14, pp. 1–32, pl. 1–32.
1911   "Die Foraminiferen (Thalamophoren) der Plankton-Expedition. Erster Teil: Die allgemeinen Organizations-verhältnisse der Foraminiferen." Plankton-Exped. Humboldt-Stiftung, Ergebn., vol. 3, pp. 1–331, pls. 1–39, text-figs. 1–110.

1949   Plate explanations for "Die Foraminiferen (Thalamophoren) der Plankton-Expedition." Micro-
            paleontologist, vol. 3, no. 2, pp. 33–40. Holograph manuscript copied and copy given to editors by
            O. W. Renz. published posthumously.

Rögl, F.
1974   "The evolution of the *Globorotolia truncatulinoides* and *Globoratolia crassaformis* groups in the
            Pliocene and Pleistocene of the Timor Trough: DSDP Leg 27, Site 262." In: Veevers, J. J., Heir-
            tzler, J. R., *et al.*, Initial Reports of the Deep Sea Drilling Project, vol. 27, pp. 743–767, pls. 1–15,
            text-figs. 1, 2.

Rögl, F. and Bolli, H. M.
1973   "Holocene to Pleistocene planktonic foraminifera of Leg 15, Site 147 (Cariaco Basin (Trench),
            Caribbean Sea) and their climatic significance." In: Edgar, N. T., Saunders, J. B. *et al.*, Initial
            Reports of the Deep Sea Drilling Project, vol. 15, pp. 553–616, pls. 1–18, text-figs. 1–6, tables 1–2.

Saito, T., Burckle, L. H. and Hays, J. D.
1975   "Late Miocene to Pleistocene biostratigraphy of Equatorial Pacific sediments." In: Saito, T. and
            Burckle, L. H. (eds.), Late Neogene Epoch Boundaries. New York: American Mus. Nat. Hist.,
            Micropaleontology Press, pp. 226–244.

Saito, T., Thompson, P. R. and Breger, D.
1976   "Skeletal ultramicrostructure of some elongate-chambered planktonic foraminifera and related
            species." In: Takayanagi, Y. and Saito, T. (eds.), Progress in Micropaleontology. New York:
            American Mus. Nat Hist., Micropaleontology Press, pp. 278–304, pls. 1–8, text-figs 1–3.

Schott, W.
1937   "Die Foraminiferen in dem äquatorialen Teil des Atlantischen Ozeans." In: Correns, C. W. "Die
            Sedimente des äquatorialen Atlantischen Ozeans." Deutsche Atlant. Exped. "Meteor" 1925–1927,
            Wiss. Ergeb., Berlin Leipzig, vol. 3, pt. 3, pp. 43–134, text-figs. 18–57.

Schubert, R. J.
1910   "Über Foraminiferen und einen Fischolithen aus dem fossilen Globigerinenschlamm von Neu-
            Guinea." Austria, Geol. Reichsanst., Verh., Vienna, pp. 318–328, pls. 1–2.
1911   "Die fossilen Foraminiferen des Bismarckarchipels und einiger angrenzender Inseln." Austria, Geol.
            Reichsanst., Abh., vol. 20, no. 4, pp. 1–130, pls. 1–6.

Schwager, C.
1866   "Fossile Foraminiferen von Kar Nikobar." Novara Exped. 1857–1859, Vienna, Bd. 2, Geol. Theil.,
            pt. 2, pp. 187–268, pls. 4–7.

Scott, G. H.
1974   "Biometry of the foraminiferal shell." In: Hedley, R. H. and Adams, C. G. (eds.), Foraminifera.
            London: Academic Press, vol. 1, pp. 55–151.

Seguenza, G.
1862   "Prime ricerche intorno ai rizopodi fossili della orgille Pleistoceniche dei dintorni di Catania."
            Accad. Gioenia Sci. Nat. Catania, Atti., Ser. 2, vol. 18, pp. 84–126, pls. 1–2.
1880   "Le formazioni terziarie nella provincia di Reggio (Calabria)." R. Accad. Lincei, Rome, Cl. Sci. Fis.,
            Mat., Nat., Mem., ser. 3, vol. 6, pp. 3–446, pls. 1–17.

Seiglie, G. A.
1963   "Una nueva especie del género *Globigerina* del Reciente de Venezuela." Oriente, Univ., Inst. O-
            ceanogr., Bol., Cumana Venezuela, 1963, vol. 2, no. 1.

Shinbo, K. and Maiya, S.
1970   "Neogene Tertiary planktonic foraminiferal zonation in the oil-producing provinces of Japan."
            Stratigraphic Correlations between Sedimentary Basins of the ECAFE Region (second volume),
            United Nations Mineral Resources Development Series, no. 36, pp. 135–142, text-figs. B 12-1, 2.

Siddall, J. D. and Brady, H. B.
1879   "Catalogue of British Recent foraminifera" Chester, England, G. R. Griffith, pp. 1–10.

Snyder, S. W.
1975   "A new Holocene species of *Globorotalia*." Tulane Stud. Geol. Paleont., vol. 11, no. 4, pp. 302–304,
            pls. 1–2.

Soldani, A.
1791   "Testaceographiae ac Zoophytographiae." Florence, vol. 1, pt. 2 (Quae reliquum secundae
            classis).

SOUTAR, A.
1971	"Micropaleontology of anaerobic sediments and the California Current." In: FUNNELL, B. M. and
RIEDEL, W. R. (eds.), The Micropalaeontology of Oceans. Cambridge Univ. Press, pp. 223–230, 1
fig., 3 tables, pls. 13.1–13.5.

SPINDLER, M., HEMLEBEN, C., BAYER, U., BÉ, A. W. H. and ANDERSON, O. R.
1979	"Lunar periodicity of reproduction in the planktonic foraminifer *Hastigerina pelagica*." Marine
Ecol. Prog. Ser., vol. 1, pp. 61–64.

SRINIVASAN, M. S. and KENNETT, J. P.
1975a	"The states of *Bolliella, Beella, Protentella* and related planktonic foraminifera based on surface
ultramicrostructure." Jour. Foram. Res., vol. 5, no. 3, pp. 155–165, pls. 1–3, text-fig. 1.
1975b	"Paleoceanographically controlled ultrastructural variation in *Neogloboquadrina pachyderma*
(EHRENBERG) at DSDP Site 284, South Pacific." In: ANDREWS, J. E., PACKHAM, G. *et al.*, Initial
Reports of the Deep Sea Drilling Project, vol. 30, pp. 709–721, pls. 1, 2.

SRINIVASAN, M. S., KENNETT, J. P. and BÉ, A. W. H.
1974	"*Globorotalia menardii neoflexuosa* new subspecies from the Northern Indian Ocean." Deep-Sea
Res., vol. 21, pp. 321–324, pl. 1.

STAINFORTH, R. M., LAMB, J. L., LUTERBACHER, H., BEARD, J. H. and JEFFORDS, R. M.
1975	"Cenozoic planktonic foraminiferal zonation and characteristics of index forms." Univ. Kansas
Paleont. Contr., art. 62, pp. 1–425, figs. 1–213.

STSCHEDRINA, Z. G.
1946	"New species of foraminifera from the Arctic Ocean." Northern Sea Route Board Drifting Expedition
on the Icebreaker "G. Sedov" in 1937–1940. Arctic Sci. Res. Inst., Trans., vol. 3 (Biology), pp. 139–
148, pls. 1–4, text-figs. 1–3. (in Russian, with English summay).

TAKAYANAGI, Y., NIITSUMA, N. and SAKAI, T.
1968	"Wall microstructure of *Globorotalia truncutulinoides* (D'ORBIGNY)." *Ibid.*, vol. 40, no. 2, pp. 141–170,
4 figs, 8 tables, pls. 20–31.

TAKAYANAGI, Y. and SAITO, T.
1962	"Planktonic foraminifera from the Nobori Formation, Shikoku, Japan." Tohoku Univ., Sci. Repts.,
2nd ser. (Geol), Spec. Vol., no. 5, pp. 67–106, pls. 24–28, text-figs. 1–2, 1 table.

TAKAYANAGI, Y., TAKAYAMA, T., SAKAI, T., ODA, M. and KATO, M.
1979	"Late Cenozoic micropaleontologic event in the Equatorial Pacific sediments." *Ibid.*, vol. 49, no. 1,
pp. 71–87. pls. 1–2.

TERQUEM, O.
1875	"Essai sur le classement des animaux qui vivent sur la plage et dans les environs de Durkergue.
Premièr fascicule." Paris, (The Author).
1876	*Op. cit.* Soc. Dunkerqunoise, Mém., Dunkerque, vol. 19 (1874–1875), pp. 405–447, pls. 1–6.

THALMANN, H. E.
1932	"Nomenclator (Um-und Neubenennungen) zu den Tafeln 1 bis 115 In: H. B. BRADY's Werk über die
Foraminiferen der Challenger Expedition, London, 1884." Eclogae Geol. Helv., vol. 25, pp. 293–
312.
1933	"Nachtrag zum Nomenclator zu BRADY's Tafelband der Foraminiferen der Challenger Expedition."
Eclogae Geol. Helv., vol. 26, pp. 251–255.

THEYER, F.
1973	"*Globorotalia inflata triangula*, a new planktonic foraminiferal subspecies." Jour. Foram. Res.,
vol. 3, no. 4, pp. 199–201, pl. 1.

THOMPSON, P. R.
1973	"Two new late Pleistocene planktonic foraminifera from a core in the southwest Indian Ocean."
Micropaleontology, vol. 9, no. 4, pp. 469–474, pls. 1–2.
1976	"Planktonic foraminiferal dissolution and the progress towards a Pleistocene equatorial Pacific
transfer function." Jour. Foram. Res., vol. 6, no. 3, pp. 208–227.
1977	"Plestocene and Recent foraminifera of the western Pacific Ocean: biostratigraphy, dissolution,
paleoecology." D. Sc. Thesis, Tohoku Univ., Sendai, Japan.
1980	"Foraminifers from Deep Sea Drilling Project Sites 434, 435, and 436, Japan Trench." In: LANG-
SETH, W., OKADA, H. *et al.*, Initial Reports of the Deep Sea Drilling Project, vols. 56–57, pt. 2,
pp. 775–808, pls. 1–9.

THOMPSON, P. R., BÉ, A. W. H., DUPLESSY, J.-C. and SHACKLETON, N. J.
1979   "Disappearance of pink-pigmented *Globigerinoides ruber* at 120,000 yr BP in the Indian and Pacific Oceans." Nature, vol. 280, no. 5723, pp. 554–558.

THOMPSON, P. R. and SAITO, T.
1974   "Pacific Pleistocene sediments: Dissolution cycles and geochronology." Geology, vol. 2, no. 7, pp. 333–335.

THOMPSON, P. R. and SCIARRILLO, J. R.
1978   "Planktonic foraminiferal biostratigraphy in the Equatorial Pacific." Nature, vol. 276, no. 5683, pp. 29–33.

THOMSON, W.
1876   "Preliminary reports to Professor WYVILLE THOMSON, F.R.S., director of the Civilian Scientific Staff, on work done on board the Challenger." Roy. Soc. London, Proc., vol. 24, pp. 426–544, pls. 20–24.

TODD, R.
1957   "Smaller foraminifera, in geology of Saipan, Mariana Island: Part. 3. Paleontology." U.S. Geol. Survey, Prof. Paper 280-H, pp. 265–320, pls. 64–93, tables 1–4.
1958   "Foraminifera from Western Mediterranean deep-sea cores." Swedish Deep-Sea Exped., Repts., vol. VIII, no. 3, pp. 167–217, tables 1–19, pls. 1–3.
1964   "Planktonic foraminifera from deep-sea cores off Eniwetok Atoll." U.S. Geol. Survey., Prof. Paper, 260–CC, pp. 1067–1100, pls. 289–295, text-figs. 319–320, tables 1–3.
1965   "The foraminifera of the tropical Pacific collections of the Albatross, 1899–1900: Part. 4. Rotaliform families and planktonic families." U.S. Nat. Mus., Bull., vol. 161, no. 4, pp. 1–139, pls. 1–28.
1967   "Planktonic foraminifera from deep-sea cores off Eniwetok Atoll." U.S. Geol. Survey, Prof. Paper 260-CC, pp. 1067–1100, pls. 289–295.

TODD, R. and BRÖNNIMANN, P.
1957   "Recent foraminifera and thecamoebina from the eastern Gulf of Paria." Cushman Found Foram. Res., Spec. Pub., no. 3, pp. 1–43, pls. 1–12.

TOLDERLUND, D. and BÉ, A. W. H.
1971   "Seasonal distribution of planktonic foraminifera in the western North Atlantic." Micropaleontology, vol. 17, no. 3, pp. 297–329.

TOLMACHOFF, I. P.
1934   "A Miocene microfauna and flora from the Atrato River, Colombia, South America." Ann. Carnegie Mus., vol. 23, pp. 275–356, pl. 41

TOWE, K. M.
1971   "Lamellar wall construction in planktonic foraminifera" In: FARINACCI, A. (ed.), Proceedings of the Second Planktonic Conference. Rome: Edizioni Tecnoscienza, vol. 2, pp. 1213–1224, pls. 1–6.

WALKER, D. A. and VILKS, G.
1973   "Spinal ultrastructure of the planktonic foraminifers *Hastigerina* THOMSON and *Globigerinella* CUSHMAN." Jour. Foram. Res., vol. 3, no. 4, pp. 196–198, pl. 1, 2 tables.

WALLER, H. O. and POLSKI, W.
1959   "Planktonic foraminifera of the Asiatic Shelf." Cushman Found. Foram. Res., Contr., vol. 10, pt. 4, pp. 123–126, pl. 10.

WIESNER, H.
1931   "Die Foraminiferen der deutschen Südpolar-Expedition 1901–1903." In: DRYGALSKI, E. VON, Deutsche Südpolar-Expedition 1901–1903. Berlin u. Leipzig: de Gruyter, Bd. 20 (Zool. Bd. 12), pp. 53–165, pls. 1–24.

WISEMAN, J. D. H. and OVEY, C. D.
1950   "Recent investigations on the deep-sea floor." Proc. Geol. Assoc., vol. 61, pt. 1, pp. 28–84, pls. 2,3.

WOOD, A.
1949   "The structure of the wall of the test in the foraminifera; its value in classification." Geol. Soc. London, Quart. Jour., vol. 104, pt. 2, pp. 229–255, pls. 13–15.

ZOBEL, B.
1968   "Phänotypische Variaten von *Globigerina dutertrei* D'ORBIGNY (Foram.); ihre Bedeutung für die Stratigraphie in quartären Tiefsee-Sedimenten." Geol. Jahrb., vol. 85, pp. 97–122, figs. 1–5, tables 1–5.